Cambridge Tracts in Mathematics and Mathematical Physics

No. 7

The Theory of Optical Instruments.

THE THEORY
OF
OPTICAL INSTRUMENTS

by

E. T. WHITTAKER, M.A., F.R.S.
Hon. Sc.D. (Dubl.); Royal Astronomer of Ireland.

CAMBRIDGE:
at the University Press
1907

CAMBRIDGE
UNIVERSITY PRESS

University Printing House, Cambridge CB2 8BS, United Kingdom

Cambridge University Press is part of the University of Cambridge.

It furthers the University's mission by disseminating knowledge in the pursuit of education, learning and research at the highest international levels of excellence.

www.cambridge.org
Information on this title: www.cambridge.org/9781107493018

First published 1907
Re-issued 2015

A catalogue record for this publication is available from the British Library

ISBN 978-1-107-49301-8 Paperback

PREFACE.

STUDENTS of Astronomy, Photography, and Spectroscopy, have frequently expressed the desire for a simple theoretical account of those defects of performance of optical instruments to which the names *coma*, *curvature of field*, *astigmatism*, *distortion*, *secondary spectrum*, *want of resolving power*, etc., are given: it is hoped that the need will to some extent be met by this little work, in which the endeavour is made to lead up directly from the first elements of Optics to those parts of the subject which are of greatest importance to workers with optical instruments. A short account of the principal instruments has been added.

While the tract is primarily written with this practical aim, the writer ventures to hope that it may be useful in drawing the attention of Pure Mathematicians to some attractive theorems: of special interest is Klein's application of the imaginary circle at infinity to establish the result (§ 30) that no optical instrument can possibly be constructed, other than the plane mirror, so as to be capable of transforming all the points of the object-space into points of the image-space.

The writer moreover believes that the customary course of Geometrical Optics presented to mathematical students in Universities might with advantage be modified: and offers the present tract as a suggestion to this end.

E. T. W.

DUNSINK OBSERVATORY, CO. DUBLIN,
November 1907.

CONTENTS.

CHAPTER I. THE POSITION AND SIZE OF THE IMAGE.

CHAPTER II. THE DEFECTS OF THE IMAGE.

CHAPTER I

THE POSITION AND SIZE OF THE IMAGE.

1. Rays and waves of light.

The existence of "shadows," which is constantly observed in every-day life, is most simply explained by the supposition that the influence to which our eyes are sensitive, and which we call *light*, travels (at any rate in air) in straight lines issuing in all directions from the "luminous" bodies with which it originates, and that it can be stopped by certain obstacles which are called *opaque*. This supposition of the *rectilinear propagation of light* is not exactly confirmed by more precise observations: light does in fact bend round the corners of opaque bodies to a certain very small extent. But the supposition is so close an approximation to the truth that it may be taken as exact without sensibly invalidating the discussion and explanation of many of the most noteworthy phenomena of light.

If an opaque screen, pierced by a small hole, be placed at some distance from a small source of light, the light transmitted through the hole will therefore travel approximately in the prolongation of the straight line joining the source to the hole. Light which is isolated in this way, so as to have approximately a common direction of propagation, is called a *pencil*: and a luminous body is to be regarded as sending out pencils of light in all directions. As there is a certain amount of vagueness in this statement, owing to the absence of any definite understanding as to what the cross-section of a pencil is to be, it is customary to make use of that principle of idealisation which is of such constant occurrence in mathematics: we introduce the term *ray* to denote a pencil whose cross-section is infinitesimally small, so that the light can be regarded as confined to a straight line: and then the above idea can be expressed by the statement that *a luminous body sends out rays of light in all directions*.

A more intimate study of the physical properties of light tends to the conviction that light consists in a disturbance of a medium which fills all space, interpenetrating material bodies: to this medium the name *aether* is given. A luminous point is then to be regarded as sending out waves of disturbance into the surrounding aether, in much the same fashion as a stone dropped into a pond sends out waves of disturbance in the water of the pond. In the latter case, we can distinguish between the *crests* of the waves, where the water is heaped up, and the *troughs*, where the surface is depressed below the normal level: these crests and troughs form a system of circles having for centre the point where the stone struck the water: we can speak of any crest or trough, or indeed any circle which has this point for centre, as a *wave-front*, meaning thereby that at all points of such a circle the water is at any instant in the same phase of disturbance. Similarly in the case of the waves emitted by a luminous point in any medium which is homogeneous (*i.e.* has the same properties at all its points) and isotropic (*i.e.* has the same properties with respect to all directions), the aether is in the same phase of disturbance at any instant at all points of a sphere having the luminous point as centre: and these surfaces of equal phase are called *wave-fronts*. It is evident that *the rays of light proceeding from the point are simply the normals to the wave-fronts.*

The luminous disturbances with which we are familiar in nature are generally of a very complicated character, but can be regarded as formed by the coexistence of a number of disturbances of simpler type, in which those wave-fronts which have the same phase (*e.g* the "crests") follow each other at regular intervals of distance. This distance is called the *wave-length* of the simple disturbance: and the time taken by one crest to move over one wave-length, *i.e.* to replace the crest in front of it, is called the *period*. Differences of wave-length or period affect the eye as differences of *colour*.

The wave-fronts are propagated outwards from a luminous point, in the same way as the water-waves on the pond: the velocity with which a wave-front moves along its own normal depends on the material medium (*e.g.* air or glass) in which the propagation is taking place. The ratio of the velocity of light *in vacuo* to the velocity in any given medium is called the *index of refraction* of the medium: it is proportional to the time light takes to travel 1 cm. in the medium. The refractive index depends to some extent on the colour of the light considered: we shall suppose for the present that we are dealing with light of some definite period, so that the index of refraction has a definite value for every medium considered.

2. Reflexion.

It is a familiar fact that light is to some extent thrown back or *reflected* from the surfaces of most bodies on which it is incident. In most cases the incident wave-front is so broken up by the small irregularities of surface of the reflecting body, that any regularity which it may have possessed before reflexion is destroyed: but if the reflecting body is capable of being used as a *mirror*, *i.e.* if its surface is optically smooth, reflexion has a regular character which we shall now investigate.

Let the plane of the diagram be perpendicular to the reflecting surface and the incident wave-front, and let AC, AB be the traces of the reflecting surface and the incident wave-front respectively. Let DC be the trace of the wave-front after reflexion, and let BC and AD be perpendicular to the respective wave-fronts, so that they are respectively parallel to the incident and reflected beams of light.

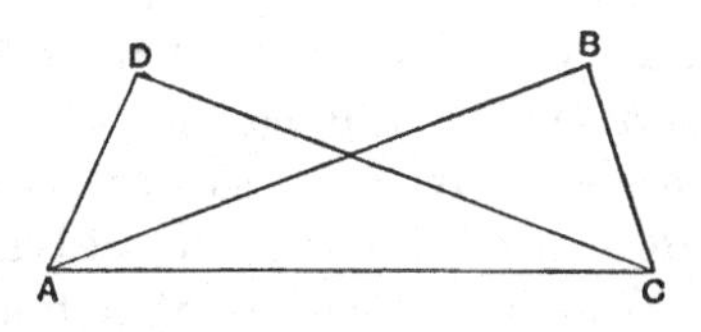

Then the time taken by the wave-front to travel from one position to the other is proportional to either BC (which represents the time taken by B to move to its new position C) or to AD (which represents the time taken by A to move to its new position D): we have therefore

$$BC = AD, \text{ or } B\hat{C}A = D\hat{A}C.$$

The angle between the incident ray BC and the normal to the surface is called the *angle of incidence*: the angle between the emergent ray AD and the normal is called the *angle of reflexion.* The last equation may be expressed by the statement that *the reflected ray is in the same plane as the incident ray and the normal to the reflecting surface, and the angle of reflexion is equal to the angle of incidence.* This is the *law of reflexion.*

3. Refraction: Fermat's principle.

If a thick piece of glass or any other transparent substance be interposed in air between a luminous body and the eye, the luminous source will in general still be seen, but will appear distorted or displaced in some manner. From this it is evident that while the rays from the luminous body which strike the glass are in part reflected at the surface of the glass, they are also partly transmitted through the glass, and at the same time experience a certain amount of deflexion from their original course. It can easily be shewn experimentally that this deflexion, to which the name *refraction* is given, takes place at the entry of the ray into the glass, and again at its emergence from the glass: there is no change of direction of the ray during its passage through the glass, if the latter be homogeneous.

If a ray of light passes from one medium into another, the acute angle between the incident ray and the normal to the interface between the media is called the *angle of incidence*, and the acute angle between the refracted ray and the normal is called the *angle of refraction*.

Refraction is easily explained as a consequence of the difference of velocity of propagation of light in different media. Let AC be the trace of a small part of the refracting surface: let AB be the trace of the incident wave-front, so that its normal BC is parallel to the incident beam: let DC be the trace of the wave-front after refraction, and AD its normal: and let μ and μ' denote the refractive indices of the media.

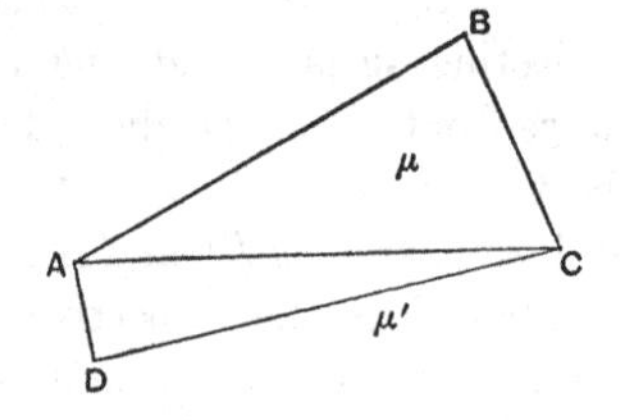

Then the time taken by the wave-front to travel from one position to the other is proportional to $\mu \,.\, BC$ (which represents the time taken by B in travelling to C) or to $\mu' \,.\, AD$ (which represents the time taken by A in travelling to D). We have therefore,

$$\mu \,.\, BC = \mu' \,.\, AD, \quad \text{or} \quad \mu \sin B\hat{A}C = \mu' \sin A\hat{C}D.$$

Thus the *law of refraction* is that *the sines of the angles of incidence and refraction are in the ratio μ'/μ.* This is readily seen to be equivalent to the statement that the cosines of the angles made by the incident and refracted rays with any line in the tangent-plane to the interface are in the ratio μ'/μ.

Media for which the index of refraction has comparatively large or small values are spoken of as *optically dense* or *optically light* respectively.

When the refraction takes place from a dense into a light medium, so that $\mu > \mu'$, the law of refraction gives a real value for the angle of refraction only when the angle of incidence is less than $\sin^{-1}(\mu'/\mu)$. This value of the angle of incidence is called the *critical angle*: when the angle of incidence is greater than the critical angle, refraction does not take place, all the light being reflected. This phenomenon is known as *total internal reflexion*.

The laws of reflexion and refraction can be comprehended in

a single statement known as *Fermat's principle*, which may be thus stated: *The path which is actually described by a ray of light between two points is such that the time taken by light in travelling from one point to the other is stationary* (*i.e.* is a maximum or minimum) *for that path as compared with adjacent paths connecting the same terminal points*: the velocity of the light being everywhere proportional inversely to the refractive index. In the case of reflexion the condition must of course be added that the path of the ray is to meet the reflecting surface.

To shew that Fermat's principle is equivalent to the ordinary law of refraction, let OA be an incident ray in a medium of index μ, AI the refracted ray in a medium of index μ', B any point near to A on the refracting surface AB. The excess of length of OB over OA is evidently $AB \cos O\hat{B}A$, and the excess of length of AI over BI is $AB \cos B\hat{A}I$: so the difference between the times of propagation of luminous disturbance along the two paths OBI and OAI is proportional to

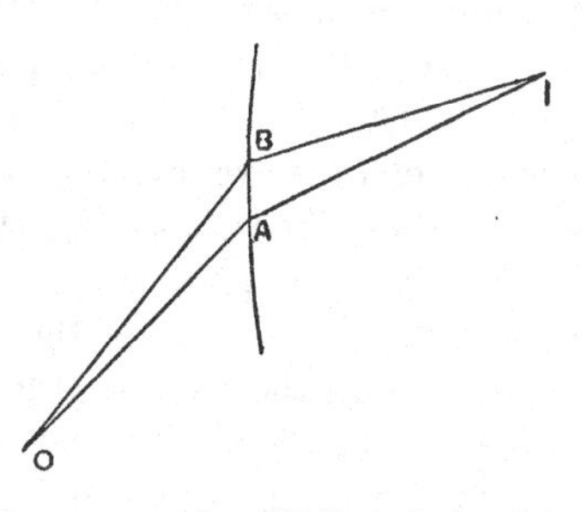

$$\mu \,.\, AB \cos O\hat{B}A - \mu' \,.\, AB \cos B\hat{A}I,$$

which vanishes in consequence of the law of refraction: this establishes the stationary property which is enunciated in Fermat's principle.

Fermat's principle is analytically expressed by the statement that

$$\int \mu ds$$

(where μ denotes the refractive index for the element ds of the path) has a stationary value, when the integration is taken along the actual path of a ray between two given terminals, as compared with adjacent curves connecting the same terminals.

4. Object and image.

In the preceding discussion of reflexion and refraction we have considered only the direction of the tangent-plane to a wave-front at some particular point: we must now proceed to consider the curvature of the wave-front, which of course depends on its distance from the luminous point from which it is diverging. The same idea can be otherwise expressed by the statement that we have hitherto treated only single *rays*, but are now about to study *pencils*.

Consider a luminous point which is emitting waves in air; we shall call this the *object-point*. Suppose that the light, after proceeding some distance from the object-point, is incident almost perpendicularly on a convex lens (*i.e.* a piece of glass bounded by two spherical faces and thickest in the middle). The waves before incidence on the lens are convex in front, so that the part of the wave-front which strikes the centre of the lens is originally a little ahead of the parts of the wave-front which strike the rim of the lens: but as the luminous disturbance travels more slowly in the glass than in air, that part of the wave which passes through the centre of the lens, and therefore has the greatest thickness of glass to traverse, will be retarded relatively to the outer parts of the wave in passing through the lens; and it may happen that this takes place to such an extent as to make the outer portions of the wave-front ahead of the central portion when the wave emerges from the lens, so that the wave is now concave in front. This concave wave will propagate itself onwards, in the direction of its own normal at every point, and thus its radius of curvature will gradually decrease until the wave finally converges to a point. This point, to which the luminous disturbance issuing from the object-point and caught by the lens is now ingathered, is said to be a *real image* of the original object-point.

In any case the centre of curvature of the wave-fronts after emergence from the lens is said to be an *image* of the object-point, the image being called *virtual* if the luminous disturbance does not actually pass through it.

5. Image-formation by direct refraction at the spherical interface between two media.

The fundamental case of image-formation is that in which the light issuing from an object is refracted at a spherical interface between two media. Let the refractive indices of the first and second media be μ and μ' respectively, and let r be the radius of curvature of the interface, counted positively when the surface is convex to the incident light. Let O be the object-point, A the *vertex* or foot of the normal from O to the interface, P a point on the interface near A, PN perpendicular to the *axis* or central line OA. We shall consider the formation of an image by a luminous disturbance which is propagated approximately along the axis.

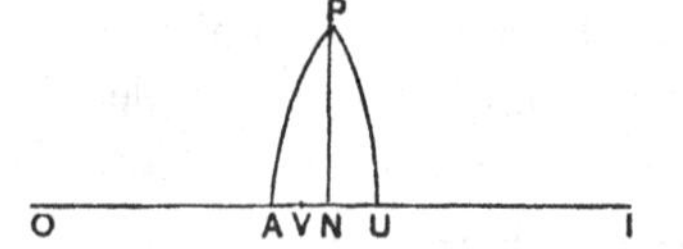

A spherical wave-front originating

from O would, but for its encounter with the second medium, occupy at some time a position represented by the trace PU, where U is a point on the axis such that $OU = OP$. But owing to the fact that the disturbance does not travel with the same velocity in the two media, the disturbance along the axis will have reached only to a point V, where

$$\mu' . AV = \mu . AU$$

or

$$(\mu' - \mu) AN - \mu . NU = \mu' . VN.$$

But by a well-known property of circles, we have

$$PN^2 = NU(ON + OU), \text{ and } PN^2 = NA(2r - NA),$$

and the equation can therefore be written in the form

$$\frac{(\mu' - \mu) . PN^2}{2r - NA} - \frac{\mu . PN^2}{ON + OU} = \mu' . VN,$$

which when P approaches indefinitely near to A takes the form

$$\frac{\mu' - \mu}{r} - \frac{\mu}{OA} = \frac{2\mu' . VN}{PN^2},$$

shewing that V and P lie on a sphere of centre I, where

$$\frac{\mu' - \mu}{r} - \frac{\mu}{OA} = \frac{\mu'}{AI}.$$

This sphere evidently represents the wave-front after refraction, and *its centre* I, determined by the last equation, *is the image-point corresponding to the object-point* O. This equation shews that the range formed by any number of object-points on the line OAI is, in the language of geometry, homographic with the range formed by the corresponding image-points.

6. Image-formation by direct refraction at any number of spherical surfaces on the same axis.

We shall consider next the successive refraction of a pencil of light at any number of spherical refracting surfaces whose centres of curvature are on the same line or *axis*; the object-point will be supposed for the present to be also situated on this axis, and the pencil of light to be directed approximately along the axis.

Let x denote the abscissa of the object-point, measured (positively in the direction of propagation of the light) from any fixed origin on the axis: and let the abscissae of the successive images be $x_1, x_2, \ldots, x'$.

Then the homographic property found in § 5 shews that x_1 is given in terms of x by an equation which can be written in the general form

$$x_1 = \frac{\alpha_1 x + \beta_1}{\gamma_1 x + \delta_1},$$

where $(\alpha_1, \beta_1, \gamma_1, \delta_1)$ are constants which depend on the position and curvature of the first refracting surface and on the refractive indices of the first and second media.

Similarly the positions of the successive images are given by equations which may be written in the form

$$x_2 = \frac{\alpha_2 x_1 + \beta_2}{\gamma_2 x_1 + \delta_2}, \qquad x_3 = \frac{\alpha_3 x_2 + \beta_3}{\gamma_3 x_2 + \delta_3}, \quad \ldots\ldots$$

Combining these so as to eliminate the intermediate images, we see that the position x' of the final image-point is determined in terms of the position x of the original object-point by an equation which can also be written in the form

$$x' = \frac{\alpha x + \beta}{\gamma x + \delta}$$

where $(\alpha, \beta, \gamma, \delta)$ are constants depending on the system of refracting surfaces, but not depending on the position of the object-point.

If γ is zero, the system is said to be a *telescopic system*: the equation which determines x' in terms x' then becomes

$$x' = \frac{\alpha}{\delta} x + \frac{\beta}{\delta},$$

which by change of origin can be written

$$x' = kx,$$

where k is a constant.

If γ is not zero (which is the more general case), we can evidently without loss of generality take γ to be unity: the equation can then be written

$$xx' + \delta x' - \alpha x - \beta = 0;$$

so if we now measure x from a point at a distance $-\delta$ from the original origin, and also measure x' from a point at a distance α from the original origin, the equation will take the form

$$xx' = C,$$

where C is a constant. *This equation determines the position x' of the final image.* The origin from which x is now measured is called the *First Principal Focus* of the optical system: it is evidently the position in which the object must be placed in order that the image may be at an infinite distance, *i.e.* in order that the emergent wave-fronts may be plane. Similarly the origin from which x' is measured is called the *Second Principal Focus*: it is the position taken by the image-point when the object-point is at an infinite distance, *e.g.* a star. In the accustomed language of geometry, the Principal Foci are the "vanishing points" of the homographic ranges formed by any set of object-points and the corresponding image-points.

7. The Helmholtz-Clausius equation.

The equation $$xx' = C$$
determines the *position* x' of the image formed by a given optical system, in terms of the position x of the object: we shall next shew how to determine the *size* of the image in terms of the size and position of the object, when the latter is supposed to be no longer a point but a body of finite (though small) dimensions.

Let AB be an object, perpendicular to the axis AA' of the instrument, and let $A'B'$ be its image; we can regard AB and $A'B'$ as two

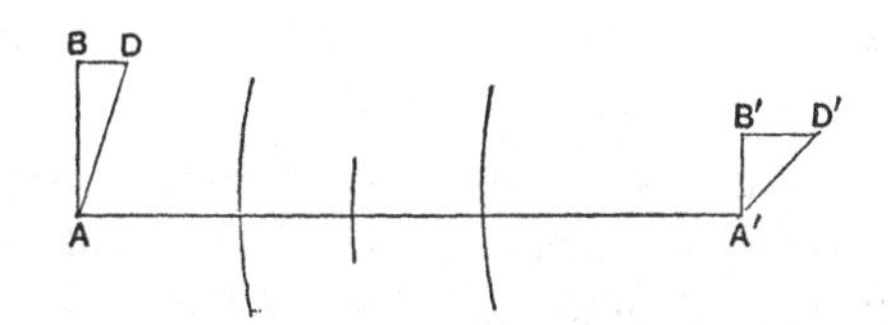

positions of a wave-front, when small quantities of the second order are neglected (the ratio of the height AB to the dimensions of the instrument being taken as a small quantity of the first order). Let AD, $A'D'$, be the corresponding two positions of another wave-front (proceeding of course from another source) slightly inclined to the first.

Then the time taken by the luminous disturbance to travel from B to B'

= " " " " " " " A to A'

= " " " " " " " D to D'.

It follows that the time taken by the light to travel the distance BD in the initial medium is equal to the time taken to travel $B'D'$ in the final medium: or

$$\mu . BD = \mu' . B'D',$$

where μ and μ' are the refractive indices of the initial and final media.

If then we denote the heights AB, $A'B'$ of the object and image by y and y' respectively, and the initial and final angles $B\hat{A}D$, $B'\hat{A}'D'$ between slightly inclined wave-fronts by α, α', respectively we have

$$\mu y \alpha = \mu' y' \alpha'.$$

This is known as *Helmholtz's equation*: it gives the *linear magnification* y'/y in terms of the *angular magnification* α'/α.

It is obvious that the above reasoning does not depend essentially on the circumstance that the optical instrument has been supposed to be symmetrical about an axis: we can therefore abandon this supposition, and state the theorem in a more general form due to Clausius*. Suppose that a small line-element l in a medium of index μ has for image a small line-element l' in a medium of index μ', and that a pencil of light which has a small angular aperture a when it issues from a point of l has an aperture a' when it converges to the corresponding image-point on l': and let ψ and ψ' be the angles made by l and l' respectively with the normals to the pencil in its plane at the two ends. Then $l \cos \psi$ will correspond to the y of Helmholtz's equation, and $l' \cos \psi'$ to y': so we obtain *Clausius' equation*

$$\mu l a \cos \psi = \mu' l' a' \cos \psi'.$$

8. The transformation of the object-space into the image-space.

We are now in a position to obtain formulae which completely determine the manner in which an optical instrument forms an image of a small object situated on its axis of symmetry.

The position of any point of a possible object, or any *point of the object-space* as it is generally called, will be specified by its abscissa x measured along the axis (positively in the direction of propagation of the light) from the First Principal Focus of the instrument, and its ordinate y drawn perpendicularly to the axis: and similarly the position of a point in the image-space will be specified by coordinates (x', y'), of which x' is measured from the Second Principal Focus of the instrument.

Suppose that two objects, of heights y_1, y_2 respectively, are at the points whose abscissae are x_1, x_2: let their images be of heights y_1', y_2', respectively. Then the equation of § 6 gives

$$xx' = C,$$

so we have

$$\text{Distance between images} = \frac{C}{x_2} - \frac{C}{x_1} = -\frac{C}{x_1 x_2} \times \text{Distance between objects}.$$

If therefore a denote the inclination to the axis of the ray from the axial point of the first object to the topmost point of the second object, and if a' denote the inclination of this ray to the axis after passing through the instrument, we have

$$\frac{a'}{a} = \frac{y_2'}{y_2} \times \frac{\text{Distance between objects}}{\text{Distance between images}}$$
$$= -\frac{y_2'}{y_2} \cdot \frac{x_1 x_2}{C}.$$

* *Ann. der Phys.* CXXI. (1864), 1.

Now if μ and μ' denote the refractive indices of the initial and final media, we have by Helmholtz's equation

$$\mu' y_1' a' = \mu y_1 a,$$

and therefore, substituting the value just found for a'/a, we have

$$-\mu' y_1' y_2' x_1 x_2 = \mu C y_1 y_2.$$

We now suppose that the two objects approach each other so as ultimately to coincide in position: thus (omitting the suffixes) we have

$$-\mu' y'^2 x^2 = \mu C y^2.$$

The equation which determines the height y' of an image in terms of the height and position of the corresponding object is therefore

$$y' = \frac{fy}{x},$$

where f is a constant connected with the constant C by the equation

$$C = -\frac{\mu'}{\mu} f^2.$$

Thus *the optical instrument transforms points* (x, y) *of the object-space near the axis into points* (x', y') *of the image-space, in a manner defined by the equations of transformation.*

$$\begin{cases} x' = -\dfrac{\mu'}{\mu}\dfrac{f^2}{x}, \\ y' = \dfrac{fy}{x}. \end{cases}$$

This transformation is of the kind called in Geometry a *collineation*, a name which is given to those transformations of space which transform points into points and also transform straight lines into straight lines.

When the initial and final media have the same refractive index, as in the case of an optical instrument in air, the above equations become

$$x' = -\frac{f^2}{x}, \quad y' = \frac{fy}{x}.$$

The constant f is called the *focal length* of the instrument. If the object is at an infinite distance (*e.g.* a pair of stars) and subtends an angle a at the instrument, it is evident from the last equation that the length of the image will be fa. Thus the focal length of a photographic telescope determines the *scale* on which the heavens will be depicted in the photographs taken with the instrument.

The object-point and image-point for which the linear magnification y'/y is unity are sometimes called the *Principal Points* of the system. The preceding equations give for the coordinates of these points

$$\frac{f}{x} = \frac{y'}{y} = 1,$$

so

$$x = f, \quad x' = -f.$$

The principal points are therefore at distances from the principal foci equal to the focal length.

9. The measure of convergence of a pencil.

When light-waves are propagated outwards from a point O in a homogeneous isotropic medium, the product of the refractive index μ and the curvature of the wave-fronts at any point P is called the *divergence* of the system of waves at the point P: the divergence is therefore measured by the quantity μ/OP.

Similarly if the luminous disturbance is converging to an image-point O, the quantity μ/PO is called the *convergence* at P. Convergence is evidently equivalent to a divergence equal in magnitude but opposite in sign.

The theorem of § 5 can thus be expressed by the statement that *the effect of direct refraction at the spherical interface (radius r) between two media μ and μ' is to increase the convergence (or diminish the divergence) of the incident pencil, by an amount $(\mu' - \mu)/r$.* This mode of stating the formula makes it easier to form a mental picture of the effect of a direct refraction on a pencil.

10. The lens.

We shall now discuss the formation of images by *lenses*. A *lens* consists of a slab of glass, or some other transparent substance, whose faces are polished, and generally spherical. The line passing through the centres of curvature of the faces is called the *axis* of the lens. We shall denote the refractive index of the material of the lens by μ, and

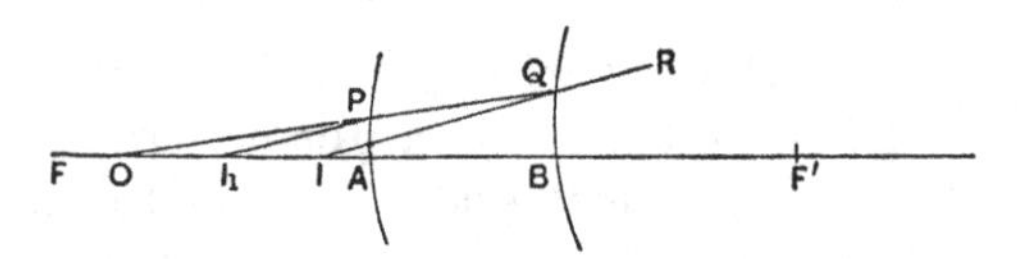

shall suppose that the lens is placed in a medium of index unity. The points A, B, in which the axis meets the faces, are called the *vertices*, and the distance AB between them is called the *thickness* of the lens, and will be denoted by t: the radii of curvature of the faces (counted positive when convex to the incident light) will be denoted by r, s; so that refraction at the first face increases the convergence of a pencil by an amount $(\mu - 1)/r$, which we shall write k_1, and refraction at the second face increases the convergence by $(1-\mu)/s$, which we shall write k_2.

Suppose that a ray OP issuing from an object-point O on the axis, and inclined at a small angle α to the axis, meets the first face of the lens at P and is refracted into the direction PQ, making an angle α_1 with the axis; and is afterwards refracted at the second face of the lens into the direction QR, making an angle α' with the axis. Let I_1 and I be the points in which the ray meets the axis after its first and second refraction respectively, so that I_1 is the place of the intermediate image of O and I is the place of the final image.

The formula of § 5, applied to the second refraction, is

$$-\frac{1}{IB} = k_2 - \frac{\mu}{I_1B},$$

or

$$\alpha' = -k_2 \,.\, BQ + \mu \,.\, \alpha_1$$

$$= -k_2 \,.\, AP + \alpha_1 (\mu - k_2 t),$$

since

$$BQ = AP + t \,.\, \alpha_1.$$

But the formula of § 5, applied to the first refraction, is

$$-\frac{\mu}{I_1A} = k_1 - \frac{1}{OA},$$

or

$$\mu\alpha_1 = -k_1 \,.\, AP + \alpha,$$

so substituting for α_1 in the preceding equation, and writing a for t/μ, we have

$$\alpha' = -k_2 \,.\, AP + (1 - k_2 a)(\alpha - k_1 \,.\, AP),$$

or

$$\frac{\alpha'}{\alpha} = -k_2 \,.\, OA + (1 - k_2 a)(1 - k_1 \,.\, OA),$$

since $AP = \alpha \,.\, OA$;

hence $\alpha'/\alpha = -K \,.\, OA + 1 - ak_2$,

where K is written for the quantity $k_1 + k_2 - ak_1k_2$.

Now Helmholtz's equation shews that $y'/y = \alpha/\alpha'$, where y'/y is the ratio of the height of the final image at I to that of the object at O. Thus we have

$$\frac{y'}{y} = \frac{1}{-K \,.\, OA + 1 - ak_2}.$$

But it was shewn in § 8 that when an image is formed by direct refraction through any optical system symmetrical about an axis, the ratio of the heights of the image and object is given by an equation of the form,

$$\frac{y'}{y} = \frac{f}{x},$$

where f is the focal length of the system and $x = FA - OA$ is the distance of the object-point from the first principal focus F, measured positively in the direction of propagation of the light. Comparing these two equations, we have

$$f = \frac{1}{K}, \qquad FA = \frac{1 - ak_2}{K}.$$

These equations determine the focal length of the lens and the position of its first principal focus; the position of the second principal focus F' is similarly given by the equation

$$BF' = \frac{1 - ak_1}{K}.$$

The position and size of the image are therefore given by the equations

$$x' = -\frac{1}{K^2 x}, \quad \text{where } F'I = x',$$

and

$$y' = \frac{y}{Kx},$$

which completely determine the image-forming action of the lens.

The distance of the vertex A from the first principal point H is (§ 8)

$$HA = FA - f = \frac{1 - ak_2}{K} - \frac{1}{K} = -\frac{ak_2}{K},$$

and the distance of the second principal point from the vertex B, measured outwards from the lens, is similarly $-ak_1/K$. The distance between the principal points is therefore $t - a(k_1 + k_2)/K$; or if t be small compared with the focal length, it is approximately $(\mu - 1)\, t/\mu$.

It is easily seen from the above formulae that generally speaking the effect of the thickness in a double-convex lens is to decrease the converging power of the lens, while in a double-concave lens the thickness increases the diverging power. When one surface of the lens is plane, the thickness has no effect on the power.

A single thick lens possesses what is known as an *optical centre*, characterised by the following property : *any ray whose direction in the glass* (*i.e.* between the two refractions) *passes through the optical centre will emerge from the second face parallel to its direction at incidence on the first face.* For if the incident ray passes through the first principal point of the lens, the emergent ray will pass through the second principal point, and by Helmholtz's equation their inclinations to the axis will be equal: so the optical centre is the image of either principal point in the corresponding face.

In the case of a lens of which one face is plane, the optical centre and one of the principal points coincide at the vertex of the curved face. In the case of a deep meniscus, *i.e.* a concavo-convex lens of great curvature, the optical centre may be at a considerable distance from the lens.

11. The thin lens.

When the lens is so thin that its thickness is negligible in comparison with its focal length, the vertices may be regarded as coincident in one point A, and the general formulae become

$$FA = AF' = \frac{1}{(\mu - 1)\left(\frac{1}{r} - \frac{1}{s}\right)} = f.$$

The principal points are now coincident at A: and the effect of the lens is simply to increase the convergence of an incident pencil by an amount $(\mu - 1)\left(\frac{1}{r} - \frac{1}{s}\right)$. This is called the *converging power* of the lens; if the lens is thicker in the middle than at the rim, it is said to be *convergent* and the focal length f is positive: if the lens is thinner in the middle than at the rim, it is said to be *divergent*, and $(\mu - 1)\left(\frac{1}{s} - \frac{1}{r}\right)$ can then be called its *diverging power*: the focal length is in this case negative. Convergent lenses form real images of objects which are situated so that the first principal focus is between the object and the lens: for the divergence of a pencil on its arrival at the lens from such an object is smaller than the converging power of the lens, so the emergent pencil converges. Divergent lenses, when used alone, cannot form real images of real objects.

When the object and image are both real (case of the convergent lens, object in front of first principal focus), or both virtual (case of the divergent lens, virtual object behind the first principal focus), the object and image are on opposite sides of the lens and the image is consequently inverted, as can be seen by observing that the straight lines joining corresponding points of the object and image cross the axis at the optical centre : in other cases the image is upright relatively to the object.

If several thin lenses are placed in contact, each lens will exercise its own converging power, and therefore the converging power of the whole is the sum of the converging powers of the separate lenses : that is, the reciprocal of the focal length of the system is the sum of the reciprocals of the focal lengths of the individual lenses.

If two thin lenses of focal lengths f_1 and f_2 are separated by an interval a, each lens will resemble a single spherical surface in converging power, and we can therefore deduce the formulae for the optical behaviour of the system from the formulae of a single thick lens, by replacing $(\mu-1)/r$ by $1/f_1$, $(1-\mu)/s$ by $1/f_2$, and t/μ by a. Thus the focal length of the system is

$$f=\frac{1}{\frac{1}{f_1}+\frac{1}{f_2}-\frac{a}{f_1 f_2}}=\frac{f_1 f_2}{f_1+f_2-a},$$

and the distance from the second lens to the second principal focus is

$$\frac{f_2(f_1-a)}{f_1+f_2-a}.$$

12. The spherical mirror.

The reflexion of a pencil of light at the spherical interface between two media can be treated in the same way as refraction. Let O be an object-point, A the foot of the normal from O to the interface, P a point on the interface near A, PN the perpendicular from P on the axis OA.

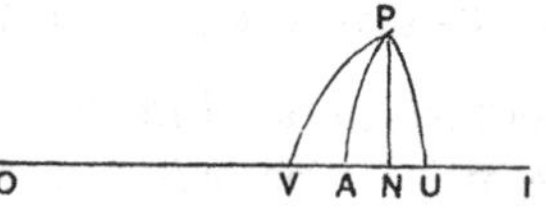

A wave-front propagated from O, which on arrival at P would have occupied the position PU if there had been no reflexion, will actually occupy the position PV, where $VA=AU$, owing to the reversal of direction of the central part of the wave by the reflexion.

Let r denote the radius of curvature of the reflecting surface, counted positively when the surface is convex to the incident light; then we have

$$VA=AU, \quad \text{or} \quad VN-AN=AN+NU.$$

Now $$PN^2 = 2r \, . \, AN,$$
since r is the radius of curvature of PA; and
$$PN^2 = 2OA \, . \, NU,$$
since OA is (in the limit) the radius of curvature of PU; and
$$PN^2 = 2AI \, . \, VN,$$
since AI is (in the limit) the radius of curvature of PV, where I denotes the image-point of O.

Thus the preceding equation becomes
$$\frac{1}{AI} - \frac{1}{r} = \frac{1}{r} + \frac{1}{OA},$$
or
$$\frac{1}{AI} = \frac{2}{r} + \frac{1}{OA}.$$

The divergence of the wave-front is therefore increased by $2/r$ *as a result of the reflexion, and the wave is at the same time reversed in direction of propagation.* The quantity $2/r$ is called the *diverging power* of the mirror.

It is easily seen from this equation that a mirror has optical properties similar to those already found for the instruments which refract light: its principal foci are coincident at the middle point of AC, where C is the centre of curvature of the mirror, and its focal length is $\frac{1}{2}r$.

13. Astigmatism.

The wave-fronts which diverge from a luminous point in a homogeneous isotropic medium are spherical. If one of these spherical wave-fronts is incident directly on an optical instrument symmetric about an axis, so that the axis of the instrument points exactly toward the luminous point, it is obvious from symmetry that the wave-front at emergence will still be symmetric about the axis, and the part of it in the immediate neighbourhood of the axis can therefore be regarded as a portion of a sphere: this is generally expressed by the statement that the emergent pencil of light is *homocentric*, a name implying that the luminous disturbance is converging to (or diverging from) a single point, namely the centre of this sphere (which of course will be the image-point of the original luminous point). If for definiteness we suppose that the pencil at emergence is converging to form a real image-point, its cross-section will gradually diminish after leaving the instrument, until at the place of the image the cross-section of the pencil reduces to a point: after this the cross-section will again increase in area; thus:

When however a thin pencil of light is incident *obliquely* on a refracting surface, the wave-front at emergence cannot in general be regarded as a portion of a sphere, for its curvature will be different in different directions along its surface: and the cross-section of the emergent pencil of light will never reduce to a single point at any distance from the instrument, but will present in succession the following forms:

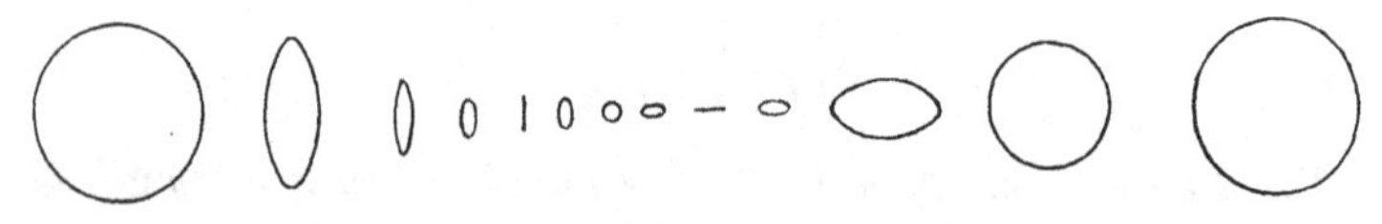

It will be observed that the cross-section reduces first to a short segment of a straight line, and subsequently to a short segment of a straight line in a direction at right angles to the first segment. These segments are called the *focal lines* of the pencil: their origin may be explained in the following way.

Let AP be the emergent wave-front, and A the point in which it is met by some ray AR_1R_2 which we select as the central or *chief ray* of the pencil: this chief ray will of course be the normal to the wave-front at A.

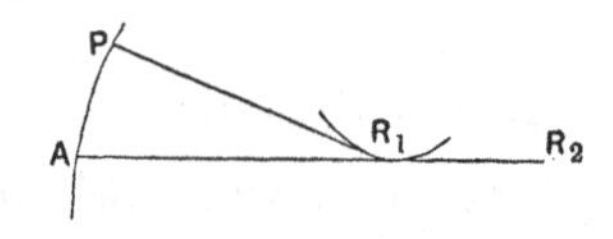

It is a well-known geometrical theorem that all the normals to a surface touch the two *caustic surfaces* which are the loci of its centres of principal curvature. Let us apply this theorem to the surface AP. Let R_1 and R_2 be the centres of principal curvature of the wave-front at A; we can suppose that the plane of the diagram is the plane of the principal section for which R_1 is the centre of curvature. Then any ray of the pencil which meets the wave-front at a point P near A touches the caustic surface through R_1 at some point near R_1, and therefore its shortest distance from the line through R_1 perpendicular to the plane of the diagram is a small quantity of at least the order AP^2/AR_1. Similarly the distance of the ray through P from the line drawn through R_2 in the plane of the diagram perpendicular to AR_2 is a small quantity of at least the order AP^2/AR_2. These lines through R_1 and R_2 are evidently the focal lines, whose existence was indicated above; R_1 and R_2 are called the *foci* of the thin pencil. We thus see that *every ray of the pencil approximately intersects the two focal lines.*

The position of the focal lines is evidently not dependent on the particular wave-front used to obtain them, since so long as the luminous disturbance remains in the same medium its wave-fronts are a family of parallel surfaces and have therefore the same caustic surfaces.

In the case of the homocentric pencils which have been considered in the theory of direct image-formation, and which are symmetrical about an axis, one caustic surface reduces to the axis itself, and the other caustic surface has near the axis the form of a surface produced by the revolution of a plane curve about a cuspidal tangent; the foci R_1 and R_2 in this case coincide at the cusp.

A thin pencil which is not homocentric, but diverges from two focal lines, is said to be *astigmatic*. If the pencil originally issued from a luminous point before the refractions, the image of this point on a screen placed at either of the foci will be a short segment of a straight line. If the screen is placed at (say) the focus R_1, the image of a *line* will therefore be quite fine and sharp if it has the same direction as the focal line at R_1, since then the short segments of lines which are the images of its individual points will overlie each other lengthways: but otherwise the image will be blurred and broad, since then the short segments which are the images of the individual points of the original line will stand out more or less perpendicularly to the general direction of the image of that line, and so will communicate breadth to it.

The theory of focal lines is really part of the general theory of *congruences*: a congruence is a set of ∞^2 lines, just as a surface is a set of ∞^2 points, and a ruled surface is a set of ∞^1 lines. Every ray of a congruence is intersected by two adjacent rays; these intersections are called the *foci* of the ray, and the two planes passing through the ray and either of its two intersecting rays are called *focal planes*. The loci of the foci are called the *focal surfaces* of the congruence: every ray of a congruence touches the focal surfaces at its focal points, and the tangent-planes are the focal planes.

If the focal planes are at right angles to each other for every ray of a congruence (as is the case in the optical application of the theory), the congruence consists of the set of normals to some surface (in the optical case, this surface is the wave-front), and is called a *normal congruence*.

14. Primary and secondary foci.

The general case of the refraction of a thin pencil of light (either homocentric or already rendered astigmatic by previous refractions) which is obliquely incident on a refracting surface of any curvature, is a somewhat complicated subject of investigation: we shall consider only the case which is of practical importance, namely the refraction of a thin pencil through an optical instrument consisting of a series of spherical refracting surfaces symmetrical about an axis, when it is assumed that the chief ray of the pencil is initially in one plane with the axis (and inclined at a finite angle to the axis), so that by symmetry the chief ray never leaves this plane in the course of the subsequent refractions. This plane through the axis and the chief ray will be called the *meridian plane* of the pencil. By symmetry it follows that the principal sections of the pencil are that by the meridian plane, which is called the *meridian* or *primary* section, and that by the plane at right angles to this, which is called the *sagittal* or *secondary* section; the corresponding foci of the pencil, which are the centres of curvature of the meridian and sagittal sections of the wave-front respectively, are called the *meridian* or *primary* focus and the *sagittal* or secondary focus. Either the meridian or the sagittal focus or any point between them, where the cross-section of the pencil is very small, may be regarded as in some sense an *image* of the object-point from which the thin pencil originally issued; but as was explained in the last article, the images thus obtained will be more or less blurred.

It is evident from symmetry that the rays which are at any time in the meridian plane of the pencil always remain in the meridian plane after any number of refractions, and that the same is true of the rays in the sagittal plane.

15. Oblique refraction of a thin pencil at a single spherical surface.

The analytical formulae for the case of a single refraction are obtained in the following way.

Let a pencil whose meridian focus is O_1 and chief ray O_1A be refracted from a medium of index μ into a medium of index μ' at a spherical interface whose centre of curvature is C and radius r, counted positive when the surface is convex to the incident light. Let AI_1 be the refracted chief ray, and let O_1PI_1 be the path of an adjacent ray in the meridian section of the pencil, so that when P is indefinitely near to A,

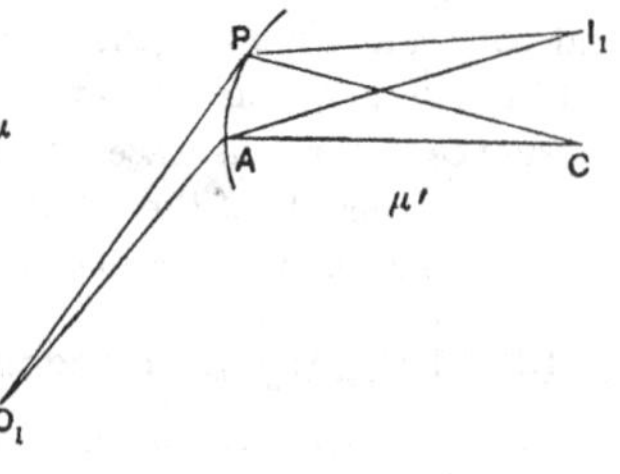

I_1 tends to a limiting position, which is that of the meridian focus of the pencil after refraction.

Let i, i' be the angles of incidence and refraction for the chief ray. Then the equation

$$\mu \sin i = \mu' \sin i'$$

when differentiated gives

$$\mu \cos i \,.\, di = \mu' \cos i' \,.\, di'$$

or
$$\mu \cos i \,(A\hat{O}_1P + A\hat{C}P) = \mu' \cos i' \,.\, (-P\hat{I}_1A + A\hat{C}P)$$

or
$$\mu \cos i \left(\frac{AP \cos i}{O_1A} + \frac{AP}{r}\right) = \mu' \cos i' \left(-\frac{AP \cos i'}{AI_1} + \frac{AP}{r}\right)$$

or
$$\frac{\mu' \cos^2 i'}{AI_1} + \frac{\mu \cos^2 i}{O_1A} = \frac{\mu' \cos i' - \mu \cos i}{r}.$$

This is the equation connecting consecutive primary foci. It may easily be interpreted geometrically as implying that the line O_1I_1 passes through a fixed point: and when i is replaced by zero it evidently reduces to the ordinary equation (§ 5) for the direct refraction of a pencil at a spherical surface.

Next, let O_2 be the sagittal focus of the incident pencil. The sagittal focus I_2 of the refracted pencil is, by symmetry, at the intersection of the chief ray AI_2 of the refracted pencil with the line of sagittal symmetry O_2C.

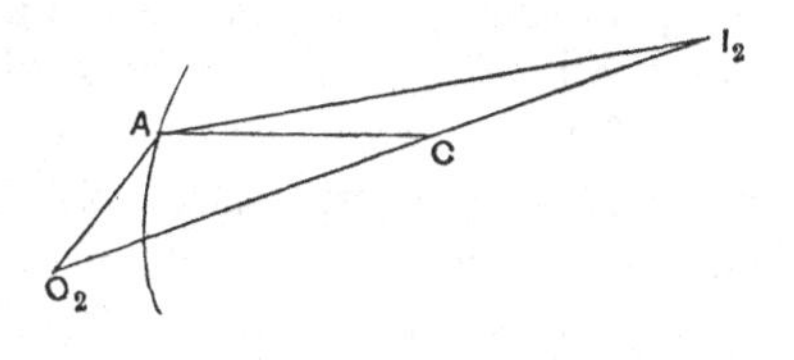

The law of refraction gives

$$\mu \sin O_2\hat{A}C = \mu' \sin C\hat{A}I_2$$

or
$$\frac{\mu \,.\, CO_2}{O_2A} = \frac{\mu' \,.\, CI_2}{AI_2}.$$

But
$$CO_2 \cos ACO_2 = O_2A \cos i + r,$$

and
$$CI_2 \cos ACO_2 = AI_2 \cos i' - r.$$

Thus we have

$$\mu \,.\, \frac{O_2A \cos i + r}{O_2A} = \mu' \,.\, \frac{AI_2 \cos i' - r}{AI_2}$$

or
$$\frac{\mu'}{AI_2} + \frac{\mu}{O_2A} = \frac{\mu' \cos i' - \mu \cos i}{r}.$$

This is the equation connecting consecutive secondary foci; like the equation for primary foci, when i is replaced by zero it reduces to the equation for direct refraction.

The union of rays at the sagittal foci is evidently, on account of the symmetry, one order higher than the union at the primary foci.

Example. A small homocentric pencil of light is incident on and reflected by a spherical surface of radius r; shew that the reflected pencil is usually astigmatic, and that the distance between the focal lines is equal to $v_1 \sim v_2$, where

$$\frac{1}{v_1} - \frac{1}{u} = \frac{2}{r\cos i}, \quad \frac{1}{v_2} - \frac{1}{u} = \frac{2\cos i}{r};$$

i being the angle of incidence and u the distance of the origin of light from the point of incidence.

16. The entrance-pupil and the field of view.

If an object is placed in front of a single convex lens, and a real image is formed behind the lens, it is obvious that of all the rays of light emitted by the object, the only ones which contribute to the formation of the image are those which pass through the lens; in other words, the cross-section of the image-forming pencils is limited by the rim of the lens. In most optical instruments the cross-sections of the image-forming pencils are limited not only by the rims of the lenses, but also by *diaphragms* or *stops*, which are generally openings in the form of circles, whose centres are on the axis of the instrument and whose planes are perpendicular to the axis; a stop evidently obstructs all those marginal rays which are at too great a distance from the axis to pass through the opening. The rims of the lenses must of course be included in an enumeration of the stops of an instrument, as also must the edge of the iris, limiting the pupil of the eye, if the instrument is used visually.

As will appear later, a judicious selection of the image-forming pencils by a suitably placed stop of small aperture may effect a great improvement in the optical performance of an instrument.

In order to find which one of the various stops in a given instrument is effective in determining the cross-section of the image-forming pencils, we consider the image of each stop formed by that part of the optical system which precedes it, and from these images we select that one which subtends the smallest angle at the axial point of the object (which may be either in front of or behind it); this image is called the *entrance-pupil.* It is evident that the cone of rays from the axial point of the object to the entrance-pupil will be able to pass through the instrument, but that a larger cone would have its marginal rays cut off by that stop of which the entrance-pupil is an image.

The angle subtended at the axial point of the object by the entrance-pupil is called the *angular aperture* of the system; the rays which proceed from the various points of the object to the axial point of the entrance-pupil are called the *chief rays* of the pencils which take part in the representation.

The image of the entrance-pupil in the entire instrument is called the *exit-pupil*: in those instruments which are intended for visual observations, the entrance-pupil of the eye should be placed at the exit-pupil of the instrument, when this is physically possible.

The stops also determine the extent of the field of view of the instrument. In order to find which one of the stops is effective in limiting the field of view, we consider the image of each stop formed by that part of the system which precedes it, and from these images we select the one which subtends the smallest angle at the axial point of the entrance-pupil: this image has been called the *entrance-window* by M. von Rohr, and evidently determines the extreme points of the object which will be represented by pencils containing chief rays; its image in the entire system is called the *exit-window*, and the angle subtended by the entrance-window at the axial point of the entrance-pupil is called the *angular field of view* of the instrument. If the entrance-window is not in the plane of the object, part of the object will be seen only by partial pencils.

17. The magnifying power of a visual instrument.

We define the *magnifying power* of a visual instrument employed to examine near objects as the ratio of the angle subtended by the image of an object at the eye, when the object is so placed that the image is at a standard distance (generally taken to be 25 cm.) from the eye, to the angle subtended by the object when viewed directly with the eye at the standard distance.

The magnifying power is therefore equal to the ratio of the heights of the image and object respectively when the image is situated at the standard distance in front of the exit-pupil of the instrument, *i.e. it is equal to the linear magnification when the image is in this position.*

When a visual instrument is used for the examination of objects at infinity, as in the case of the astronomical telescope, it is natural to define the magnifying power as the angular magnification at the pupils: this by Helmholtz's theorem (§ 7) is equal to the reciprocal of the linear magnification at the pupils, so *the magnifying power is equal to the ratio of the radius of the entrance-pupil to the radius of the exit-pupil.*

CHAPTER II.

THE DEFECTS OF THE IMAGE.

18. The removal of astigmatism from an optical instrument with a narrow stop.

We now proceed to consider the conditions which must be satisfied in order that an optical instrument may, as accurately as possible, transform pencils issuing from the various points of the object into homocentric pencils in the image-space, so that the image may be a point-for-point representation of the object without blurring: and moreover, that the image so formed may be geometrically similar to the object.

It will be supposed throughout that we are dealing with an object at some definite distance from the instrument, and that we wish to eliminate errors in the image for an object in this position alone: if the object is moved to some other position, errors will of course reappear in the image. It will therefore be assumed that a plane object is placed at right angles to the axis of the instrument: and we shall suppose at first that a diaphragm of very small aperture is placed at some point on the axis, so that the pencils of light which pass through it, and by which alone the image is formed, are of very small cross-section. Under these assumptions we shall find the condition which must be satisfied in order that these pencils when they finally emerge into the image-space may be homocentric, *i.e.* that the image may be free from astigmatism. The treatment will necessarily be approximate, the linear dimensions of the object and of the lens-apertures being supposed as in Chapter I to be small compared with the radii of curvature of the refracting surfaces; but the approximation is now to be carried to a higher order than in Chapter I.

Let the ith refracting surface be taken to separate a medium of index μ_{i-1} from a medium of index μ_i, and to have a radius of curvature r_i, measured positively when the surface is convex to the incident light; let l_{i-1} denote the height of the intermediate image of the object before refraction at this surface, and l_i the height of the intermediate image after this refraction: let x_i and x_i' be the distances of the intermediate images of the diaphragm from this refracting surface before and after this refraction respectively (distances being measured positively in the direction of propagation of the light), and s_i and s_i' the distances of the intermediate images of the object from the surface before and after this refraction; and let i and i' be the angles of incidence and refraction at this surface for the chief ray of the pencil proceeding from the topmost point of the object.

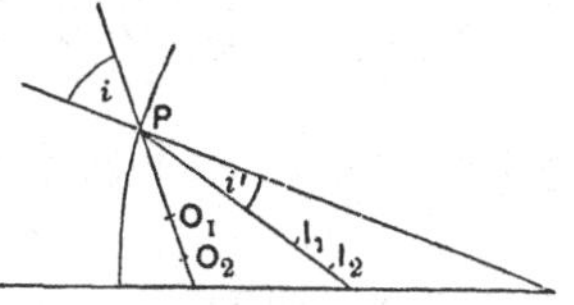

Then if O_1 and O_2 are the primary and secondary foci of this pencil before its refraction at the rth surface at P, and I_1, I_2 are the primary and secondary foci after this refraction, we have (§ 15)

$$\frac{\mu_i \cos^2 i'}{PI_1} - \frac{\mu_{i-1}\cos^2 i}{PO_1} = \frac{\mu_i \cos i' - \mu_{i-1}\cos i}{r_i} = \frac{\mu_i}{PI_2} - \frac{\mu_{i-1}}{PO_2}.$$

Since, to our degree of approximation, we have

$$\cos^2 i = 1 - i^2, \quad \cos^2 i' = 1 - i'^2,$$

these equations give

$$\frac{\mu_i}{PI_1} - \frac{\mu_i i'^2}{PI_1} - \frac{\mu_{i-1}}{PO_1} + \frac{\mu_{i-1} i^2}{PO_1} = \frac{\mu_i}{PI_1} - \frac{\mu_i \,.\, I_1 I_2}{PI_1^2} - \frac{\mu_{i-1}}{PO_1} + \frac{\mu_{i-1}\,.\,O_1O_2}{PO_1^2},$$

or

$$\frac{\mu_i \,.\, I_1 I_2}{s_i'^2} - \frac{\mu_{i-1}\,.\,O_1O_2}{s_i^2} = \frac{\mu_i \,.\, i'^2}{s_i'} - \frac{\mu_{i-1}\,.\,i^2}{s_i}.$$

Now if y_i denote the distance of P from the axis, we have

$$i = \frac{y_i}{x_i} - \frac{y_i}{r_i}, \qquad i' = \frac{y_i}{x_i'} - \frac{y_i}{r_i},$$

and we have (§ 5)

$$\mu_{i-1}\left(\frac{1}{r_i} - \frac{1}{s_i}\right) = \mu_i\left(\frac{1}{r_i} - \frac{1}{s_i'}\right) = Q_{si} \text{ say,}$$

and

$$\mu_{i-1}\left(\frac{1}{r_i} - \frac{1}{x_i}\right) = \mu_i\left(\frac{1}{r_i} - \frac{1}{x_i'}\right) = Q_{xi} \text{ say,}$$

so our equation can be written

$$\frac{\mu_i \,.\, I_1 I_2}{s_i'^2} - \frac{\mu_{i-1} \,.\, O_1 O_2}{s_i^2} = y_i^2\, Q_{xi}^2 \left(\frac{1}{\mu_i s_i'} - \frac{1}{\mu_{i-1} s_i}\right).$$

But by similar triangles we have

$$\frac{l_{i-1}}{x_i - s_i} = \frac{y_i}{x_i},$$

and we have

$$Q_{xi} - Q_{si} = \mu_{i-1}\left(\frac{1}{s_i} - \frac{1}{x_i}\right),$$

so

$$Q_{xi} - Q_{si} = \frac{\mu_{i-1} l_{i-1}}{s_i y_i}.$$

Thus the equation becomes

$$\frac{\mu_i \,.\, I_1 I_2}{s_i'^2} - \frac{\mu_{i-1} \,.\, O_1 O_2}{s_i^2} = \frac{\mu_{i-1}^2 l_{i-1}^2}{s_i^2}\left(\frac{Q_{xi}}{Q_{xi} - Q_{si}}\right)^2 \left(\frac{1}{\mu_i s_i'} - \frac{1}{\mu_{i-1} s_i}\right).$$

Since by Helmholtz's theorem we have

$$\mu_i s_i l_i = \mu_{i-1} s_i' l_{i-1},$$

this can be written

$$\frac{I_1 I_2}{\mu_i l_i^2} - \frac{O_1 O_2}{\mu_{i-1} l_{i-1}^2} = \left(\frac{Q_{xi}}{Q_{xi} - Q_{si}}\right)^2 \left(\frac{1}{\mu_i s_i'} - \frac{1}{\mu_{i-1} s_i}\right).$$

Now add together the equations of this type for all the refracting surfaces in the instrument. The only terms surviving on the left-hand side will be one involving the $O_1 O_2$ of the original object and one involving the $I_1 I_2$ of the final image: but the former of these vanishes, since the pencils issuing from the object are originally homocentric: and the latter term must vanish if the pencils converging to the final image are also to be homocentric. Thus we have the theorem that *the condition for absence of astigmatism in the final image is*

$$\sum_i \left(\frac{Q_{xi}}{Q_{xi} - Q_{si}}\right)^2 \left(\frac{1}{\mu_i s_i'} - \frac{1}{\mu_{i-1} s_i}\right) = 0,$$

where the summation is taken over all the refracting surfaces. This is known as *Zinken-Sommer's condition.*

19. The removal of astigmatism from an optical instrument used at full aperture.

If an optical instrument can be constructed so as to give emergent pencils which are free from astigmatism even when a narrow diaphragm is not inserted, *i.e.* when the full aperture of the lenses is filled by the pencils, it is evident that the emergent wave-fronts will have their principal radii of curvature equal at every point, and will therefore be spherical: that is, the emergent wave-fronts will converge to points, and the instrument will furnish an image which corresponds point for point with the object. Clearly if this absence of astigmatism for full pencils is to be attained, the condition found in the last article must be satisfied independently of the diaphragm: in other words, the last equation must be true whatever value x_i may have. We shall now find the conditions which must be satisfied in order that this may be the case.

If h_i denotes the height at which a paraxial ray (*i.e.* a ray whose path lies indefinitely close to the axis), passing through the axial points of the intermediate images, meets the ith refracting surface, and if d_{i-1} denotes the distance between the $i-1$th and ith refracting surfaces, we evidently have (the other symbols being defined as in the last article)

$$x'_{i-1} = x_i + d_{i-1}, \quad s'_{i-1} = s_i + d_{i-1}, \quad s'_{i-1}/h_{i-1} = s_i/h_i,$$

so

$$d_{i-1} = \frac{d_{i-1}(x_i - s_i)}{x_i - s_i} = \frac{x_i s'_{i-1} - s_i x'_{i-1}}{x_i - s_i}$$

$$= \frac{x_i s'_{i-1}}{x_i - s_i} - \frac{s_i x'_{i-1}}{x'_{i-1} - s'_{i-1}}$$

$$= \frac{s'_{i-1}}{s_i\left(\dfrac{1}{s_i} - \dfrac{1}{x_i}\right)} - \frac{s_i}{s'_{i-1}\left(\dfrac{1}{s'_{i-1}} - \dfrac{1}{x'_{i-1}}\right)}$$

$$= \frac{\mu_{i-1} h_{i-1}}{h_i (Q_{xi} - Q_{si})} - \frac{\mu_{i-1} h_i}{h_{i-1}(Q_{x,\,i-1} - Q_{s,\,i-1})},$$

or

$$\frac{1}{h_i^2 (Q_{xi} - Q_{si})} - \frac{1}{h^2_{i-1}(Q_{x,\,i-1} - Q_{s,\,i-1})} = \frac{d_{i-1}}{\mu_{i-1} h_{i-1} h_i}.$$

Adding together equations of this type, we have

$$\frac{1}{h_i^2 (Q_{xi} - Q_{si})} = \sum_{p=1}^{p=i-1} \frac{d_p}{\mu_p h_p h_{p+1}} + \frac{1}{h_1^2 (Q_{x1} - Q_{s1})}.$$

Now the condition found in the last article for absence of astigmatism with a narrow diaphragm at x_1 is

$$\sum_i \left(\frac{Q_{xi}}{Q_{xi} - Q_{si}}\right)^2 \left(\frac{1}{\mu_i s_i'} - \frac{1}{\mu_{i-1} s_i}\right) = 0,$$

or

$$\sum_i \left(1 + \frac{Q_{si}}{Q_{xi} - Q_{si}}\right)^2 \left(\frac{1}{\mu_i s_i'} - \frac{1}{\mu_{i-1} s_i}\right) = 0,$$

and by use of the preceding equation this can be written

$$\sum_i \left\{1 + Q_{si} h_i^2 \sum_{p=1}^{i-1} \frac{d_p}{\mu_p h_p h_{p+1}} + \frac{Q_{si} h_i^2}{h_1^2 (Q_{x1} - Q_{s1})}\right\}^2 \left(\frac{1}{\mu_i s_i'} - \frac{1}{\mu_{i-1} s_i}\right) = 0 \ldots \text{(A)}.$$

The only quantity in this equation which involves the position of the diaphragm is the quantity Q_{x1}; so the equation will be satisfied for *all* positions of the diaphragm, provided the coefficients of the various powers of $\frac{1}{Q_{x1}-Q_{s1}}$ are separately zero; that is, *the optical instrument will give point-images when used at full aperture, provided it satisfies the three conditions*

$$\sum_i Q_{si}^2 h_i^4 \left(\frac{1}{\mu_i s_i'} - \frac{1}{\mu_{i-1} s_i}\right) = 0 \ldots\ldots \text{(I)},$$

$$\sum_i \left\{1 + Q_{si} h_i^2 \sum_{p=1}^{i-1} \frac{d_p}{\mu_p h_p h_{p+1}}\right\} Q_{si} h_i^2 \left(\frac{1}{\mu_i s_i'} - \frac{1}{\mu_{i-1} s_i}\right) = 0 \ \ldots \text{(II)},$$

$$\sum_i \left\{1 + Q_{si} h_i^2 \sum_{p=1}^{i-1} \frac{d_p}{\mu_p h_p h_{p+1}}\right\}^2 \left(\frac{1}{\mu_i s_i'} - \frac{1}{\mu_{i-1} s_i}\right) = 0 \ldots \text{(III)}.$$

These are known as *Seidel's first, second, and third conditions, respectively**. We shall now proceed to interpret them.

20. Seidel's first condition: the removal of spherical aberration.

We shall first interpret Seidel's condition (I).

By comparing condition (I) with equation (A) of the last article, it is evident that condition (I) taken alone represents the condition that the instrument shall give point-images by all pencils which can pass through a diaphragm specified by the condition $Q_{x1} - Q_{s1} = 0$, *i.e.* subject to the presence of a narrow stop placed at the axial point of the object. But a narrow stop placed at the axial point of the object would allow the passage of a full pencil from this axial point, while it would not allow any light whatever to reach the instrument from the other points of the object. *Condition (I) therefore implies that all rays proceeding from the axial point of the object are accurately united into the axial point of the image.* This is usually expressed by the statement that *the optical instrument has no spherical aberration.*

When condition (I) is not satisfied, the rays proceeding from the axial point of the object do not reunite to form a single image-point; the *marginal* rays, or rays which pass through the outer zones† of the lenses, do not meet the axis in the same point as the paraxial rays. When the instrument forms a real image, if the image as formed

* Seidel, *Astr. Nach.* XLIII. col. 289.

† The term *zone* is used to denote a ring-shaped region of one of the refracting surfaces, bounded by two circles whose centres are on the axis.

by the marginal rays is nearer to the instrument than that formed by the paraxial rays, the instrument is said to be *under-corrected* for spherical aberration : in the opposite case, it is said to be *over-corrected.*

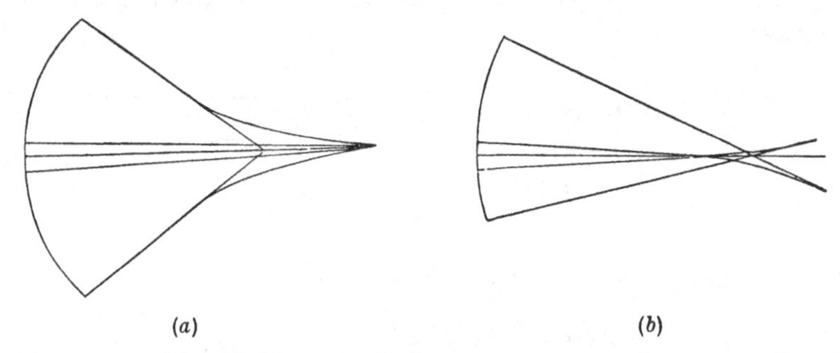

The figures (*a*) and (*b*) respectively represent an under-corrected and an over-corrected pencil.

The curve drawn in the figures, touched by the rays of the pencil, is the *caustic* (§ 13) : it is the evolute of the wave-front. If the light is received on a screen placed nearer to the instrument than the focus of an under-corrected pencil, the image will evidently be surrounded by a hard edge (where the caustic meets the screen) : but if the screen is placed beyond the focus, the image will be surrounded by scattered light.

Spherical aberration will evidently become more and more noticeable as the cross-section of the pencil increases, *i.e.* as the aperture of the optical system increases.

21. Evaluation of the spherical aberration in uncorrected instruments.

When the spherical aberration is not eliminated in an optical instrument, its amount can be determined in the following way.

Let A be the vertex of the ith refracting surface AP, and let O be the intersection of the axis with the prolongation of an image-forming

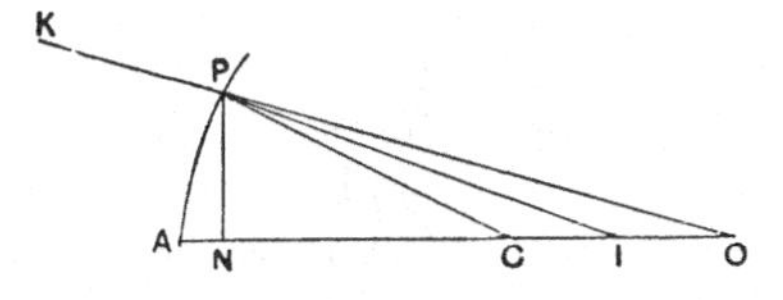

ray KP in the $(i-1)$th medium, while I is the intersection of the axis with the same ray PI in the ith medium. Denote AO by $s_i + \Delta_{i-1}$, and

AI by $s_i' + \Delta_i$, so that Δ_i measures the spherical aberration in the ith medium for a ray which meets the ith refracting surface at a height $PN = h_i$.

Then if C be the centre and r_i the radius of the refracting surface, we have

$$\tan A\hat{O}P = \frac{h_i}{NO} = \frac{h_i}{NC + AO - r_i} = \frac{h_i}{r_i\left(1 - \frac{h_i^2}{r_i^2}\right)^{\frac{1}{2}} + s_i + \Delta_{i-1} - r_i}$$

$$= \frac{h_i}{s_i}\left(1 + \frac{1}{2}\frac{h_i^2}{r_i s_i} - \frac{\Delta_{i-1}}{s_i}\right),$$

to our degree of approximation; whence we readily have

$$\sin A\hat{O}P = \frac{h_i}{s_i}\left\{1 - \frac{\Delta_{i-1}}{s_i} + \frac{h_i^2}{2s_i}\left(\frac{1}{r_i} - \frac{1}{s_i}\right)\right\},$$

and

$$\cos A\hat{O}P = 1 - \frac{h_i^2}{2s_i^2}.$$

Similarly

$$\tan A\hat{I}P = \frac{h_i}{s_i'}\left(1 + \frac{1}{2}\frac{h_i^2}{r_i s_i'} - \frac{\Delta_i}{s_i'}\right),$$

$$\sin A\hat{I}P = \frac{h_i}{s_i'}\left\{1 - \frac{\Delta_i}{s_i'} + \frac{h_i^2}{2s_i'}\left(\frac{1}{r_i} - \frac{1}{s_i'}\right)\right\},$$

$$\cos A\hat{I}P = 1 - \frac{h_i^2}{2s_i'^2},$$

$$\sin A\hat{C}P = \frac{h_i}{r_i}, \quad \cos A\hat{C}P = 1 - \frac{h_i^2}{2r_i^2}.$$

Now by the law of refraction we have

$$\mu_{i-1}\sin(A\hat{C}P - A\hat{O}P) = \mu_i \sin(A\hat{C}P - A\hat{I}P).$$

Substituting for the sines and cosines in this equation their values just found, we have

$$\mu_{i-1}\left(\frac{1}{r_i} - \frac{1}{s_i} + \frac{\Delta_{i-1}}{s_i^2}\right) + \frac{\mu_{i-1}h_i^2}{2s_i}\left(\frac{1}{r_i} - \frac{1}{s_i}\right)^2$$
$$= \mu_i\left(\frac{1}{r_i} - \frac{1}{s_i'} + \frac{\Delta_i}{s_i'^2}\right) + \frac{\mu_i h_i^2}{2s_i'}\left(\frac{1}{r_i} - \frac{1}{s_i'}\right)^2,$$

or

$$\frac{\mu_i\Delta_i}{s_i'^2} - \frac{\mu_{i-1}\Delta_{i-1}}{s_i^2} = \tfrac{1}{2}Q_{si}^2 h_i^2\left(\frac{1}{\mu_{i-1}s_i} - \frac{1}{\mu_i s_i'}\right).$$

Now if θ_i denotes the inclination of the ray to the axis in the ith medium, we have

$$\theta_{i-1}s_i = \theta_i s_i' = h_i,$$

so the equation becomes

$$\mu_i\Delta_i\theta_i^2 - \mu_{i-1}\Delta_{i-1}\theta^2_{i-1} = \tfrac{1}{2}Q^2_{si}h_i^4\left(\frac{1}{\mu_{i-1}s_i} - \frac{1}{\mu_i s_i'}\right).$$

Adding together the equations of this type which refer to the successive refracting surfaces, we have

$$\mu_i\Delta_i\theta_i^2 = \tfrac{1}{2}\sum_{p=1}^{p=i} Q^2_{sp}h_p^4\left(\frac{1}{\mu_{p-1}s_p} - \frac{1}{\mu_p s_p'}\right),$$

so

$$\Delta_i = \frac{s_i'^2}{2\mu_i h_i^2}\sum_{p=1}^{i} Q^2_{sp}h_p^4\left(\frac{1}{\mu_{p-1}s_p} - \frac{1}{\mu_p s'_p}\right).$$

This equation gives the spherical aberration of the image at any stage.

If we apply the formula to the case of image-formation by a single thin lens, of refractive index μ, radii r and s, and focal length f, so that

$$\frac{1}{f} = (\mu - 1)\left(\frac{1}{r} - \frac{1}{s}\right),$$

we have for the spherical aberration, along the axis, of a ray incident at height h and proceeding from an object at infinity (*e.g.* a star) the expression

$$\Delta = \tfrac{1}{2}f^2h^2\left\{-\frac{Q_1^2}{\mu s_1'} + Q_2^2\left(\frac{1}{\mu s_1'} - \frac{1}{f}\right)\right\}.$$

But $$Q_1 = \frac{1}{r},\quad Q_2 = \frac{1}{s} - \frac{1}{f} = \frac{1}{r} - \frac{\mu}{(\mu-1)f},\quad \frac{\mu}{s_1'} = \frac{\mu - 1}{r},$$

and substituting these values we have

$$\Delta = \tfrac{1}{2}fh^2\left\{-\frac{2}{\mu r^2} + \frac{1}{(\mu-1)fr} - \left(\frac{1}{r} - \frac{\mu}{(\mu-1)f}\right)^2\right\}.$$

By applying this equation to particular cases it will be found, for example, that a plano-convex lens is strongly under-corrected when the plane face is turned towards an object at infinity, but only feebly under-corrected when the convex face is turned towards the object.

The spherical aberration of a lens can however be completely changed, and brought to any desired value, if in the process of polishing the faces of the lens are made to depart from the exact spherical form. If for example we consider a telescope objective which is affected by spherical aberration, so that the longitudinal aberration of a ray at distance h from the axis (the object being supposed at infinity) is βh^2, it can without difficulty be shewn that this aberration can be completely removed by figuring the inner face so as to remove a film of glass whose thickness at the point h is a constant $+\frac{\beta h^4}{4(\mu-1)f^2}$, where f is the focal length of the objective and μ the index of the glass on the inner side.

22. Coma and its removal: the Fraunhofer condition.

We next proceed to the interpretation of Seidel's second condition. If we write equation (A) of § 19 in the form

$$X + Y(Q_{x1} - Q_{s1}) + Z(Q_{x1} - Q_{s1})^2 = 0 \quad \text{...........(A)},$$

the three Seidel conditions are respectively

$$X = 0, \quad Y = 0, \quad Z = 0.$$

Now equation (A) represents the condition which must be satisfied in order that astigmatism may be absent for that position of the stop which corresponds to the value of $(Q_{x1} - Q_{s1})$ in the equation: and we can regard the above form of the equation as a Taylor series developing the condition in ascending powers of $(Q_{x1} - Q_{s1})$. The vanishing of X implies (§ 20) the absence of astigmatism when $(Q_{x1} - Q_{s1})$ is zero, *i.e.* when the stop is exactly at the axial point of the object: similarly the vanishing of X and Y together implies that the astigmatism is not only zero when $Q_{x1} - Q_{s1}$ is zero, but that its rate of increase is zero when $Q_{x1} - Q_{s1}$ is made slightly different from zero, *i.e.* when the stop is placed slightly in front of the object but very near to it. But when the stop is in this position, it will permit the passage of practically the full pencil from any point of the object which is very near the axis: and hence the full pencil from such a point will be free from astigmatism on emergence from the instrument. In this way we see that *when Seidel's condition (II) is satisfied in addition to condition (I), there is a point-for-point representation not only of the axial point of the object, but also of points of the object which are infinitesimally near the axis.*

The defect of the image which is thus removed may be further elucidated in the following way.

Suppose that the instrument does not satisfy condition (II), and consider the full meridian pencil from an object-point O situated just off the axis. The rays on emergence from the instrument will touch a caustic ABC (Fig. a). If the light be received on a screen BK at right angles to the axis at the place of the image, it is evident that no light will reach the screen above the point B, where the caustic meets the screen. The rays which have passed through the central zone of the instrument will meet the screen at B in a bright point (B in Figs. a and b). The rays which have passed through a zone of the instrument somewhat further from the centre will (as is evident

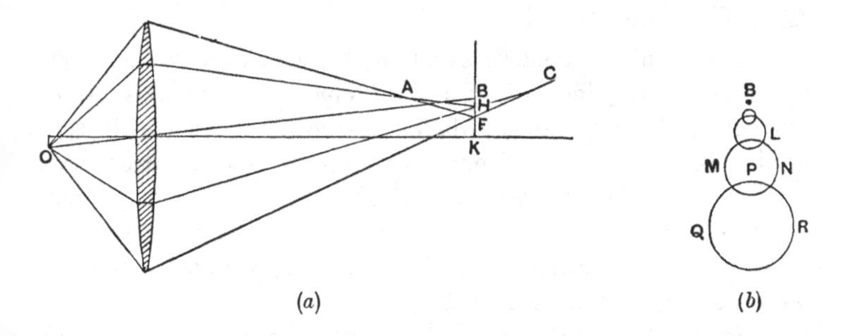

(*a*) (*b*)

from Fig. a) meet the screen lower down than B (at H in Fig. a) in a circular section (LMN in Fig. b): and the rays which have passed through the outermost zones of the instrument will meet the screen still lower down, in a still larger circle (F in Fig. a, PQR in Fig. b). In this way we see that the total effect on the screen is a balloon-shaped flare of light, bright at the tip B and growing fainter as it expands downwards*. This defect is known as *coma* (κόμη, the hair): it is of great importance, as *e.g.* the definition in the outer parts of the field of an astronomical telescope (assuming good definition at the centre of the field) depends chiefly on the removal of coma. It is perhaps more difficult to grasp than any of the other defects, owing probably to the bewildering variety of (at first sight) unrelated ways in which it may be described: from one point of view we may regard it as spherical aberration (of the primary focus) for object-points just off the axis: from another point of view we may regard it as implying that the linear magnification of a very small object, situated on the axis of the instrument, is different when different zones of the instrument are used to form the image. To our order of approximation, and on the assumption that there is no spherical aberration for the axial point of the image, these two statements are evidently equivalent.

The condition (II) for the removal of coma was called by Seidel *Fraunhofer's condition*, because it was found to be almost exactly

* It is to be observed that each point of a circle such as PQR in the coma corresponds to *two* diametrically opposite points of the zone which gives rise to the circle, *e.g.* it is evident from Fig. a that the two extreme marginal rays will meet the screen in the *same* point F: one-half of a zone gives a whole comatic circle.

satisfied by the Königsberg Heliometer objective, which had been constructed by Fraunhofer many years before the discovery of the condition, and which was celebrated for the excellence of its definition.

23. The sine condition.

We have seen that Seidel's equation (II) expresses the condition that the linear magnification of a small object on the axis of the instrument shall be the same whatever zone of the lenses is used in forming the image. In all our work hitherto, however, it has been assumed that the fourth power (and higher powers) of the angular aperture can be neglected: and we shall now shew that the condition just stated can be expressed analytically in a form which is rigorous however large the aperture may be.

Suppose then that the lenses of an optical instrument are of any size; and let O be a small object situated on the axis in a medium of index μ, its height l being at right angles to the plane of the diagram.

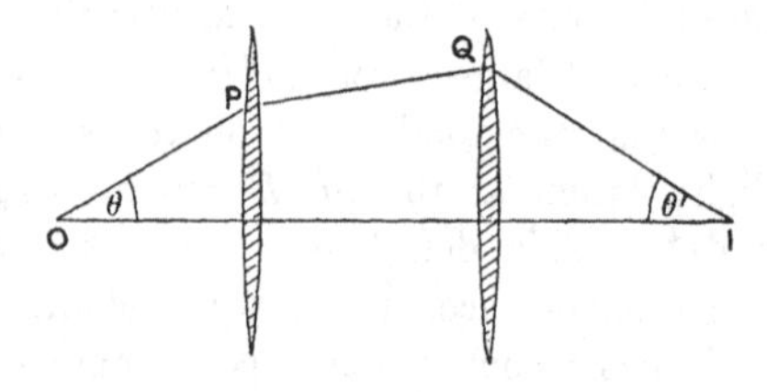

Let the instrument form an image I of O, in a medium of index μ', by a thin sagittal pencil whose plane is at right angles to the plane of the diagram, and whose chief ray $OPQI$ makes an angle θ with the axis initially, and θ' finally. Let α denote the angle between the extreme rays of the pencil initially, and let α' be the final value of this angle: and suppose that $d\phi$ is the angle between the meridian planes which pass through the extreme rays of the pencil, so

$$\alpha = \sin\theta \,.\, d\phi, \quad \alpha' = \sin\theta' \,.\, d\phi.$$

Clausius' equation (§ 7) gives at once

$$\mu \alpha l = \mu' \alpha' l',$$

or

$$\mu \sin\theta \,.\, l = \mu' \sin\theta' \,.\, l',$$

so *the linear magnification of a small object, when the image is formed by rays which pass through this zone on the refracting surfaces, is*

$$\frac{\mu \sin\theta}{\mu' \sin\theta'}.$$

This result is true for all optical instruments, independently of whether they are affected with spherical aberration or not.

Suppose now that the instrument is corrected for spherical aberration, so that the images of O formed by different zones are situated at the same point of the axis. In order that the images of a small object at O may be in all respects identical, they must be of the same size; and therefore the equation

$$\frac{\mu \sin \theta}{\mu' \sin \theta'} = m,$$

where m is the linear magnification for the image formed by the paraxial rays, must be satisfied by every ray which issues from the axial point O. This equation is called the *sine-condition.*

As might be expected, the sine-condition also ensures that the images formed by meridian pencils have the same magnification, whatever be the zones through which the pencils pass. For again applying Clausius' equation (§ 7)

$$\mu \cos \psi \,.\, la = \mu' \cos \psi' \,.\, l'a',$$

we have in this case (the object and image being taken in the plane of the diagram, perpendicular to the axis)

$$\psi = \theta, \quad \psi' = \theta', \quad a = d\theta, \quad a' = d\theta',$$

so the equation becomes

$$\mu \cos \theta d\theta \,.\, l = \mu' \cos \theta' \,.\, d\theta' \,.\, l'.$$

But by differentiating the sine-condition we have

$$\mu \cos \theta \, d\theta = m\mu' \cos \theta' \, d\theta',$$

so

$$l'/l = m,$$

i.e. the magnification is m whatever zone of the lenses is employed.

The honour of discovering the sine-condition must be shared between Seidel*, who first gave that approximate form of it which he called Fraunhofer's condition, and Clausius†, who first obtained the rigorous form. It remained unnoticed however until in 1873 it was rediscovered by Abbe and Helmholtz.

24. Aplanatism.

If an optical instrument is free from spherical aberration, and also satisfies the sine-condition, for a certain position of the object, it is said to be *aplanatic* for the object in question.

* *Astr. Nach.* XLIII. (1856), 289.

† *Pogg. Ann.* CXXI (1864), 1.

3—2

In the construction of microscope objectives, use is made of the fact that there is one position of the object for which a single spherical refracting surface is aplanatic: a result which we shall now proceed to establish.

Let C be the centre of a sphere of glass of radius r and of index μ, situated in a medium of index unity: suppose that an object O is embedded in the glass at a distance CO equal to r/μ from the centre; and let I be the point on CO at a distance μr from C.

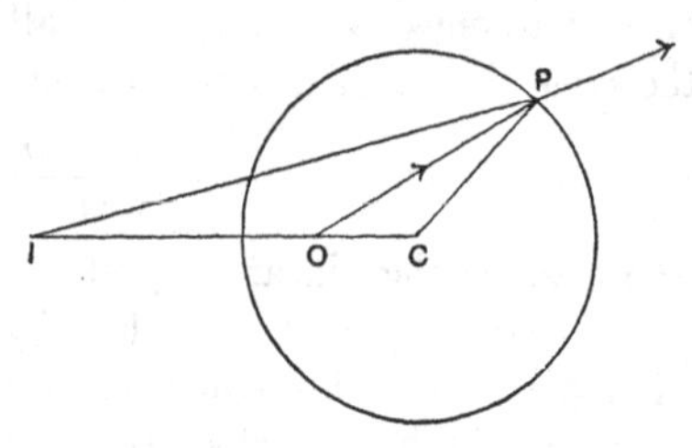

Then if P be any point on the spherical surface, we have

$$OC/CP = PC/CI,$$

so the triangles OCP, PCI, are similar: and therefore we have

$$\frac{\sin I\hat{P}C}{\sin O\hat{P}C} = \frac{\sin P\hat{O}C}{\sin O\hat{P}C} = \frac{PC}{OC} = \mu.$$

This shews that a ray proceeding from O in the direction OP will be refracted at the surface exactly into the direction IP, whether P is near the axis IOC or not: in other words, there is no spherical aberration for the positions O and I of the object and image.

But it is also true that the sine-condition is satisfied for this position of the object: for we have

$$\frac{\mu \sin P\hat{O}C}{\sin P\hat{I}C} = \frac{\mu \sin P\hat{O}C}{\sin O\hat{P}C} = \mu \,.\, \frac{PC}{OC} = \mu^2,$$

shewing that the linear magnification is independent of the zone of the spherical surface at which the refraction takes place, and is equal to μ^2. *The spherical surface is therefore aplanatic for an object in the position O.* The application of this principle to microscopes will be discussed later.

There is another well-known case in which spherical aberration is perfectly corrected for pencils of any aperture, namely that in which the rays of light from a star are received on a concave reflecting surface having the form of a paraboloid of revolution whose axis is directed toward the star. In this case, as is obvious from the geometry of the paraboloid, the rays are accurately united into an image at the focus of the paraboloid: but it can readily be verified that in this case the sine-condition is not satisfied, so the surface is not aplanatic. It is this want of aplanatism which causes the deterioration of definition in the outer parts of the field of a reflecting telescope.

25. Derivation of the Fraunhofer condition from the sine-condition.

We shall now shew analytically (what has already become obvious from general reasoning) that the Fraunhofer condition for absence of coma is simply the approximate form of the sine-condition, when the fourth and higher powers of the angular aperture are neglected.

The sine-condition is (§ 23)

$$\frac{\sin\theta'}{\sin\theta} = \frac{\mu}{\mu' m},$$

where m is the linear magnification for the paraxial rays.

Now considering separately the refraction at the ith refracting

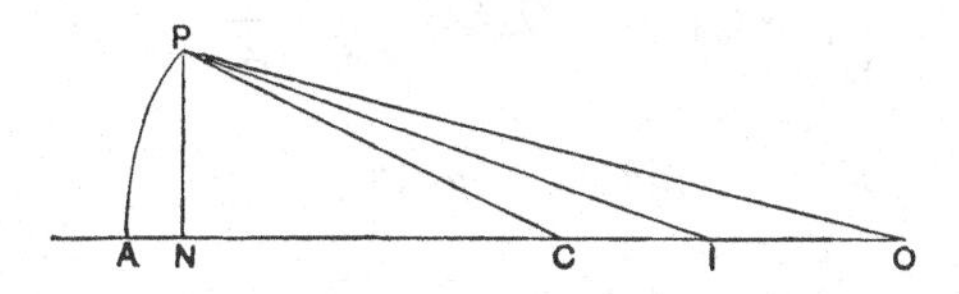

surface, and using notation similar to that which has been frequently used before, we have

$$\frac{\sin\theta_i'}{\sin\theta_i} = \frac{PO}{PI} = \frac{(NO^2+y^2)^{\frac{1}{2}}}{(NI^2+y^2)^{\frac{1}{2}}} = \frac{NO + \dfrac{y^2}{2NO}}{NI + \dfrac{y^2}{2NI}},$$

or

$$\frac{\sin\theta_i'}{\sin\theta_i} = \frac{AO - \dfrac{h_i^2}{2r_i} + \dfrac{h_i^2}{2s_i}}{AI - \dfrac{h_i^2}{2r_i} + \dfrac{h_i^2}{2s_i'}}.$$

Consequently we have

$\dfrac{\sin\theta'}{\sin\theta}$ = the product of the values of $\dfrac{\sin\theta_i'}{\sin\theta_i}$ for the separate refractions

$$= \text{the product of the quantities } \frac{s_i}{s_i'}\cdot\left(\frac{AO}{AI}\cdot\frac{s_i'}{s_i}\right)\cdot\frac{1-\dfrac{h_i^2}{2s_i}\left(\dfrac{1}{r_i}-\dfrac{1}{s_i}\right)}{1-\dfrac{h_i^2}{2s_i'}\left(\dfrac{1}{r_i}-\dfrac{1}{s_i'}\right)}.$$

It will be observed that AO differs from s_i by the spherical aberration Δ_{i-1}.

Now the product of the quantities $\frac{s_i}{s_i'}$ is $\frac{\mu}{\mu' m}$: so if the sine-condition is satisfied, we must have

$$\text{product of quantities } \frac{\frac{s_i+\Delta_{i-1}}{s_i}}{\frac{s_i'+\Delta_i}{s_i'}} \frac{1-\frac{h_i^2}{2s_i}\left(\frac{1}{r_i}-\frac{1}{s_i}\right)}{1-\frac{h_i^2}{2s_i'}\left(\frac{1}{r_i}-\frac{1}{s_i'}\right)}=1,$$

or
$$\sum_i \left\{\frac{\Delta_{i-1}}{s_i}-\frac{\Delta_i}{s_i'}-\frac{Q_{si}h_i^2}{2\mu_{i-1}s_i}+\frac{Q_{si}h_i^2}{2\mu_i s_i'}\right\}=0,$$

where the summation is taken over the various refracting surfaces.

Substituting for Δ_{i-1} and Δ_i from § 21, this becomes

$$\sum_i\left\{-\frac{1}{\mu_{i-1}\theta^2_{i-1}s_i}\sum_{p=1}^{i-1} Q^2_{sp}h_p^4\left(-\frac{1}{\mu_{p-1}s_p}+\frac{1}{\mu_p s_p'}\right)\right.$$
$$\left.+\frac{1}{\mu_i\theta_i^2 s_i'}\sum_{p=1}^{i} Q^2_{sp}h_p^4\left(-\frac{1}{\mu_{p-1}s_p}+\frac{1}{\mu_p s_p'}\right)+Q_{si}h_i^2\left(\frac{1}{\mu_i s_i'}-\frac{1}{\mu_{i-1}s_i}\right)\right\}=0,$$

or
$$\sum_i (1+Q_{si}h_i^2 A_i)\, Q_{si}h_i^2\left(\frac{1}{\mu_i s_i'}-\frac{1}{\mu_{i-1}s_i}\right)=0,$$

where
$$A_i=\frac{1}{\mu_i\theta_i^2 s_i'}-\frac{1}{\mu_i\theta_i^2 s_{i+1}}+\frac{1}{\mu_{i+1}\theta^2_{i+1}s'_{i+1}}-\frac{1}{\mu_{i+1}\theta^2_{i+1}s_{i+2}}+\dots$$
$$=\frac{1}{\mu_i\theta_i h_i}-\frac{1}{\mu_i\theta_i h_{i+1}}+\frac{1}{\mu_{i+1}\theta_{i+1}h_{i+i}}-\dots$$
$$=-\left(\frac{d_i}{\mu_i h_i h_{i+1}}+\frac{d_{i+1}}{\mu_i h_{i+1}h_{i+2}}+\dots\right).$$

The sine-condition thus becomes

$$\sum_i\left(1-Q_{si}h_i^2\sum_{p=i}^{n}\frac{d_p}{\mu_p h_p h_{p+1}}\right)Q_{si}h_i^2\left(\frac{1}{\mu_i s_i'}-\frac{1}{\mu_{i-1}s_i}\right)=0.$$

But Seidel's condition (I), which is supposed to be satisfied, is (§ 19)

$$\sum_i Q^2_{si}h_i^4\left(\frac{1}{\mu_i s_i'}-\frac{1}{\mu_{i-1}s_i}\right)=0.$$

Multiplying the latter equation by

$$\sum_{p=1}^{n}\frac{d_p}{\mu_p h_p h_{p+1}},$$

and adding it to the former, we have

$$\sum_i\left\{1+Q_{si}h_i^2\sum_{p=1}^{i-1}\frac{d_p}{\mu_p h_p h_{p+1}}\right\}Q_{si}h_i^2\left(\frac{1}{\mu_i s_i'}-\frac{1}{\mu_{i-1}s_i}\right)=0,$$

and this is no other than the Fraunhofer condition already found in §§ 19, 22.

26. Astigmatism and Seidel's third condition.

Of Seidel's three conditions (§ 19), only the third now remains for interpretation. Since the three conditions together ensure freedom from astigmatism over the whole field, and the two first conditions have been shewn to relate specially to the central parts of the field, it is evident that *when Seidel's two first equations are satisfied, the third equation may be regarded as representing the condition for removal of astigmatism from the outer parts of the field.*

27. Petzval's condition for flatness of field.

We have seen that the wave-fronts which issue from points of the object will, after passage through an optical instrument, converge again to points forming an image, provided that, in instruments with very narrow diaphragms, the Zinken-Sommer condition (§ 18) is satisfied; or, in instruments for which the diaphragm is not narrow, provided Seidel's equations (I), (II), (III) (§ 19) are satisfied. It remains to consider whether this image is a faithful copy of the object.

A condition which must obviously be satisfied if this is to be the case is that if the object is plane and at right angles to the axis, the image shall also be plane; by symmetry, it will also be at right angles to the axis. We shall now find the analytical equation which must be satisfied by the lenses of the instrument in order that a plane object may give a plane image; it is usually referred to as the *condition for flatness of field.*

Let AP be the ith refracting surface, O_0O the intermediate image before refraction at this surface, I_0I the image after refraction at this surface, PO and PI the directions of the chief ray (§ 16) of the pencil by which the image-points O and I are formed; and let X and X' be the intermediate images of the diaphragm.

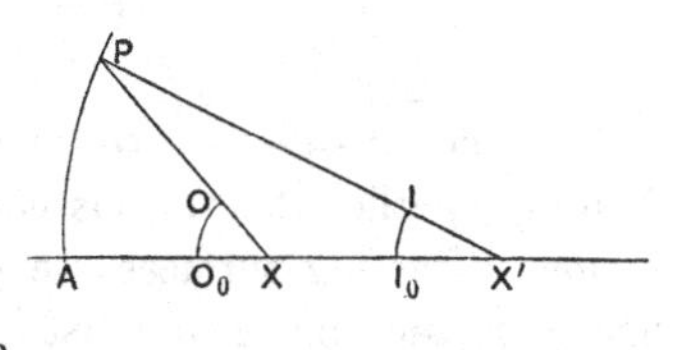

Let the radii of curvature of O_0O and I_0I respectively be ρ_{i-1} and ρ_i; and let the notation in other respects be the same as in previous articles.

Then the coordinates of O referred to the vertex A are $\left(s_i + \frac{l^2_{i-1}}{2\rho_{i-1}},\ l_{i-1}\right)$; those of I are $\left(s_i' + \frac{l_i^2}{2\rho_i},\ l_i\right)$, and those of P are $\left(\frac{y_i^2}{2r_i},\ y_i\right)$.

We have therefore

$$PI = \left\{\left(s_i' + \frac{l_i^2}{2\rho_i} - \frac{y_i^2}{2r_i}\right)^2 + (l_i - y_i)^2\right\}^{\frac{1}{2}},$$

so

$$\frac{1}{PI} = \frac{1}{s_i'}\left\{1 - \frac{l_i^2}{2\rho_i s_i'} + \frac{y_i^2}{2r_i s_i'} - \frac{(l_i - y_i)^2}{2s_i'^2}\right\}$$

to our approximation, and similarly

$$\frac{1}{PO} = \frac{1}{s_i}\left\{1 - \frac{l^2_{i-1}}{2\rho_{i-1}s_i} + \frac{y_i^2}{2r_i s_i} - \frac{(l_{i-1} - y_i)^2}{2s_i^2}\right\}.$$

Thus the equation (§ 15)

$$\frac{\mu_i \cos i' - \mu_{i-1}\cos i}{r_i} = \frac{\mu_i}{PI} - \frac{\mu_{i-1}}{PO}$$

becomes

$$\frac{\mu_i - \mu_{i-1}}{r_i} - \frac{\mu_i i'^2 - \mu_{i-1} i^2}{2r_i} = \frac{\mu_i}{s_i'}\left\{1 - \frac{l_i^2}{2\rho_i s_i'} + \frac{y_i^2}{2r_i s_i'} - \frac{(l_i - y_i)^2}{2s_i'^2}\right\}$$
$$- \frac{\mu_{i-1}}{s_i}\left\{1 - \frac{l^2_{i-1}}{2\rho_{i-1}s_i} + \frac{y_i^2}{2r_i s_i} - \frac{(l_{i-1} - y_i)^2}{2s_i^2}\right\}.$$

Since

$$i' = -\frac{Q_{xi}y_i}{\mu_i}, \qquad i = -\frac{y_i Q_{xi}}{\mu_{i-1}},$$

$$l_{i-1} = \frac{y_i s_i}{\mu_{i-1}}(Q_{xi} - Q_{si}), \quad l_i = \frac{y_i s_i'}{\mu_i}(Q_{xi} - Q_{si}),$$

this equation becomes

$$\frac{Q_{xi}^2}{r_i}\left(\frac{1}{\mu_{i-1}} - \frac{1}{\mu_i}\right) = -\frac{(Q_{xi} - Q_{si})^2}{\mu_i\rho_i} + \frac{\mu_i}{r_i s_i'^2} - \frac{\mu_i}{s_i'^3}\left\{\frac{s_i'}{\mu_i}(Q_{xi} - Q_{si}) - 1\right\}^2$$
$$+ \frac{(Q_{xi} - Q_{si})^2}{\mu_{i-1}\rho_{i-1}} - \frac{\mu_{i-1}}{r_i s_i^2} + \frac{\mu_{i-1}}{s_i^3}\left\{\frac{s_i}{\mu_{i-1}}(Q_{xi} - Q_{si}) - 1\right\}^2,$$

or

$$\frac{1}{\mu_i\rho_i} - \frac{1}{\mu_{i-1}\rho_{i-1}} = \left(\frac{Q_{xi}}{Q_{xi} - Q_{si}}\right)^2\left(\frac{1}{\mu_{i-1}s_i} - \frac{1}{\mu_i s_i'}\right) + \frac{1}{r_i}\left(\frac{1}{\mu_i} - \frac{1}{\mu_{i-1}}\right).$$

Adding together the various equations of this type which refer to the various refracting surfaces, we see that if the original object and final image are each plane we must have

$$\sum_i \left\{\left(\frac{Q_{xi}}{Q_{xi} - Q_{si}}\right)^2\left(\frac{1}{\mu_{i-1}s_i} - \frac{1}{\mu_i s_i'}\right) + \frac{1}{r_i}\left(\frac{1}{\mu_i} - \frac{1}{\mu_{i-1}}\right)\right\} = 0.$$

The first sum is however known to be zero, since the instrument satisfies Zinken-Sommer's condition (§ 18): and hence we see that *the condition for flatness of field is*

$$\sum_i \frac{1}{r_i}\left(\frac{1}{\mu_i} - \frac{1}{\mu_{i-1}}\right) = 0.$$

This condition was first given by Petzval, and is known by his name.

If the instrument consists of a number of thin lenses in air, the refractive index and focal length of the kth lens being μ_k and f_k respectively, the condition obviously becomes

$$\sum_k \frac{1}{\mu_k f_k} = 0.$$

It is interesting to observe that the Petzval condition does not depend in any way on the distance of the object from the instrument, or on the separation of its component lenses.

28. The condition for absence of distortion.

Having now secured flatness of field, it remains to ensure that the object (supposed to be a plane figure at right angles to the axis of the instrument) shall give rise to an image which is geometrically similar to itself. When this is not the case, the image is said to be affected by *distortion.*

Distortion, in an optical instrument symmetrical about an axis, simply means that the magnification of the image is not the same in the outer parts of the field as at the centre. When the magnification is greatest at the centre, a straight line in the outer part of the object-field will evidently give rise to an image-line which is curved, with its concavity turned towards the centre of the field: this is known as "barrel" distortion. If on the other hand the magnification is greatest at the margin of the field, a straight line in the outer part of the object-field will give rise to a curved line in the image-field, with its convexity turned towards the centre of the field: this is known as "pin-cushion" distortion. All single lenses, whether consisting of one lens or of several lenses cemented together, produce distortion: it is therefore necessary for most purposes in Photography to use objectives in which there are one or more intervals between the lenses.

We shall use the same notation as in the preceding articles; and

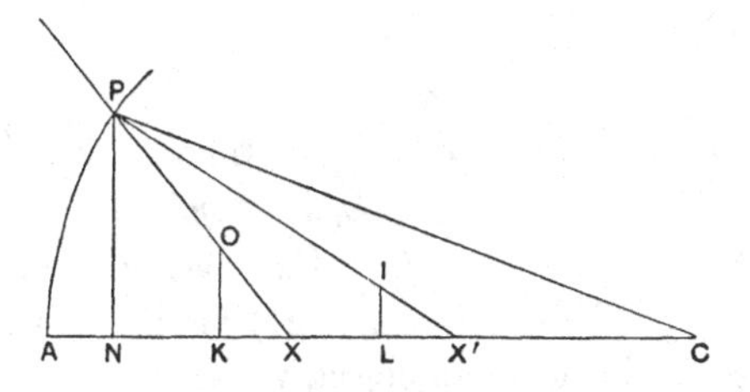

shall in addition denote by ϕ_i the angle which the chief ray of the image-forming pencil of a point I of the intermediate image makes with the axis of the instrument in the medium μ_i; and we shall denote by $x_i' + E_i$ the distance from the ith refracting surface AP of the intermediate image X' of the diaphragm, formed by this pencil in the medium μ_i, so that E_i really represents the spherical aberration of this image of the diaphragm.

The distance of the intermediate image O from the axis before this refraction is OK (where OK is the perpendicular to the axis from O), or $KX \tan\phi_{i-1}$, or $(x_i + E_{i-1} - s_i)\tan\phi_{i-1}$; and its height after this refraction is $IL = (x_i' + E_i - s_i')\tan\phi_i$. If there is no distortion, the product of the ratios IL/OK at the various refracting surfaces must be independent of the position of the point O in the object; so the product

$$\prod_i \frac{(x_i' + E_i - s_i')\tan\phi_i}{(x_i + E_{i-1} - s_i)\tan\phi_{i-1}}$$

must be independent of the height y_i at which the chief ray PI meets the ith refracting surface.

Now
$$\frac{\tan\phi_i}{\tan\phi_{i-1}} = \frac{NX}{NX'} = \frac{AX - \dfrac{y_i^2}{2r_i}}{AX' - \dfrac{y_i^2}{2r_i}} = \frac{x_i + E_{i-1} - \dfrac{y_i^2}{2r_i}}{x_i' + E_i - \dfrac{y_i^2}{2r_i}},$$

so the product in question is

$$\prod_i \frac{(x_i' - s_i')\left(1 + \dfrac{E_i}{x_i' - s_i'}\right) x_i \left(1 + \dfrac{E_{i-1}}{x_i} - \dfrac{y_i^2}{2r_i x_i}\right)}{(x_i - s_i)\left(1 + \dfrac{E_{i-1}}{x_i - s_i}\right) x_i' \left(1 + \dfrac{E_i}{x_i'} - \dfrac{y_i^2}{2r_i x_i'}\right)}.$$

Neglecting factors which do not depend on y_i, this product is

$$\prod_i \frac{\left(1 + \dfrac{E_i}{x_i' - s_i'}\right)\left(1 + \dfrac{E_{i-1}}{x_i} - \dfrac{y_i^2}{2r_i x_i}\right)}{\left(1 + \dfrac{E_{i-1}}{x_i - s_i}\right)\left(1 + \dfrac{E_i}{x_i'} - \dfrac{y_i^2}{2r_i x_i'}\right)};$$

as this reduces to unity for paraxial rays, it must be always unity: we must therefore have

$$\sum_i \left(\frac{E_i}{x_i' - s_i'} + \frac{E_{i-1}}{x_i} - \frac{y_i^2}{2r_i x_i} - \frac{E_{i-1}}{x_i - s_i} - \frac{E_i}{x_i'} + \frac{y_i^2}{2r_i x_i'}\right) = 0,$$

or
$$\sum_i \left\{\frac{\mu_i E_i}{x_i'^2 (Q_{xi} - Q_{si})} - \frac{\mu_{i-1} E_{i-1}}{x_i^2 (Q_{xi} - Q_{si})} + \frac{y_i^2}{2r_i}\left(\frac{1}{x_i'} - \frac{1}{x_i}\right)\right\} = 0.$$

But applying to E_i the formula (§ 21) for the spherical aberration of the intermediate image of an object, we have

$$\frac{\mu_i E_i}{x_i'^2} - \frac{\mu_{i-1} E_{i-1}}{x_i^2} = \tfrac{1}{2} Q^2_{xi} y_i^2 \left(\frac{1}{\mu_{i-1} x_i} - \frac{1}{\mu_i x_i'}\right).$$

The condition for absence of distortion is therefore

$$\sum_i y_i^2 \left\{ \frac{Q_{xi}^2}{Q_{xi} - Q_{si}} \left(\frac{1}{\mu_{i-1} x_i} - \frac{1}{\mu_i x_i'} \right) + \frac{1}{r_i} \left(\frac{1}{x_i'} - \frac{1}{x_i} \right) \right\} = 0.$$

Now by Helmholtz's theorem, $\mu_{i-1} l_{i-1} \theta_{i-1}$ is a constant for all the images; but $\theta_{i-1} = h_i / s_i$, so

$$\frac{\mu_{i-1} l_{i-1}}{s_i} = \frac{\text{constant}}{h_i};$$

and since (§ 18) we have

$$y_i = \frac{\mu_{i-1} l_{i-1}}{s_i (Q_{xi} - Q_{si})},$$

we see that

$$y_i = \frac{\text{constant}}{h_i (Q_{xi} - Q_{si})}.$$

The condition for absence of distortion may therefore be written

$$\sum_i \frac{1}{h_i^2 (Q_{xi} - Q_{si})^2} \left\{ \frac{Q^2_{xi}}{Q_{xi} - Q_{si}} \left(\frac{1}{\mu_{i-1} x_i} - \frac{1}{\mu_i x_i'} \right) + \frac{1}{r_i} \left(\frac{1}{x_i'} - \frac{1}{x_i} \right) \right\} = 0.$$

Writing

$$\frac{1}{r_i} - \frac{Q_{xi}}{\mu_{i-1}} \text{ for } \frac{1}{x_i}, \quad \text{and} \quad \frac{1}{r_i} - \frac{Q_{xi}}{\mu_i} \text{ for } \frac{1}{x_i'},$$

this becomes

$$\sum_i \frac{Q_{xi}}{h_i^2 (Q_{xi} - Q_{si})^3} \left\{ Q^2_{xi} \left(\frac{1}{\mu_i^2} - \frac{1}{\mu^2_{i-1}} \right) + \left(\frac{1}{\mu_{i-1}} - \frac{1}{\mu_i} \right) \frac{2Q_{xi} - Q_{si}}{r_i} \right\} = 0,$$

or

$$\sum_i \frac{Q_{xi}}{h_i^2 Q_{si} (Q_{xi} - Q_{si})^3} \left\{ Q_{si} Q^2_{xi} \left(-\frac{1}{\mu^2_{i-1}} + \frac{1}{\mu_i^2} \right) + \frac{(Q_{si} - Q_{xi})^2}{r_i} \left(-\frac{1}{\mu_{i-1}} + \frac{1}{\mu_i} \right) - \frac{Q_{xi}^2}{r_i} \left(-\frac{1}{\mu_{i-1}} + \frac{1}{\mu_i} \right) \right\} = 0,$$

or

$$\sum_i \frac{Q_{xi}}{h_i^2 Q_{si} (Q_{xi} - Q_{si})^3} \left\{ Q_{xi}^2 \left(\frac{1}{\mu_{i-1} s_i} - \frac{1}{\mu_i s_i'} \right) + \frac{(Q_{si} - Q_{xi})^2}{r_i} \left(\frac{1}{\mu_i} - \frac{1}{\mu_{i-1}} \right) \right\} = 0,$$

or

$$\sum_i \frac{Q_{xi}}{h_i^2 Q_{si} (Q_{xi} - Q_{si})} \left\{ \frac{Q_{xi}^2}{(Q_{xi} - Q_{si})^2} \left(\frac{1}{\mu_{i-1} s_i} - \frac{1}{\mu_i s_i'} \right) + \frac{1}{r_i} \left(\frac{1}{\mu_i} - \frac{1}{\mu_{i-1}} \right) \right\} = 0.$$

This is the required condition for absence of distortion, when the diaphragm is at the position x; it being assumed that the instrument already satisfies the Zinken-Sommer condition and the Petzval condition.

If it is required that distortion should be absent when the image is formed by pencils filling the whole aperture of the optical instrument, we must find the condition in order that the last equation may be satisfied whatever value x may have; it being now assumed that the instrument already satisfies the three Seidel conditions of § 19, and the Petzval condition (§ 27). For this purpose we substitute for $\frac{Q_{xi}}{Q_{xi} - Q_{si}}$ its value (§ 19)

$$1 + h_i^2 Q_{si} \left\{ \sum_{p=1}^{i-1} \frac{d_p}{\mu_p h_p h_{p+1}} + \frac{1}{h_1^2 (Q_{x1} - Q_{s1})} \right\};$$

making this substitution, and omitting terms which vanish in consequence of the conditions already satisfied, the condition for absence of distortion becomes

$$\sum_i \frac{1}{h_i^2 Q_{si}} \left\{ 1 + h_i^2 Q_{si} \sum_{p=1}^{i-1} \frac{d_p}{\mu_p h_p h_{p+1}} \right\} \left[\left\{ 1 + h_i^2 Q_{si} \sum_{p=1}^{i-1} \frac{d_p}{\mu_p h_p h_{p+1}} \right\}^2 \left(\frac{1}{\mu_{i-1} s_i} - \frac{1}{\mu_i s_i'} \right) + \frac{1}{r_i} \left(\frac{1}{\mu_i} - \frac{1}{\mu_{i-1}} \right) \right] = 0.$$

This does not involve the position of the diaphragm, so is *the required condition for absence of distortion with full pencils.*

If we denote

$$- \frac{1}{Q_{si} h_i^2} \left\{ 1 + Q_{si} h_i^2 \sum_{p=1}^{i-1} \frac{d_p}{\mu_p h_p h_{p+1}} \right\} \text{ by } U_i,$$

and

$$Q_{si}^2 h_i^4 \left(\frac{1}{\mu_{i-1} s_i} - \frac{1}{\mu_i s_i'} \right) \text{ by } \Theta_i,$$

we see, on collating the results of the preceding articles, that the condition for absence of

spherical aberration is $\Sigma \Theta_i = 0$,

coma ,, $\Sigma \Theta_i U_i = 0$,

astigmatism ,, $\Sigma \Theta_i U_i^2 = 0$,

curvature of field ,, $\Sigma \frac{1}{r_i} \left(\frac{1}{\mu_i} - \frac{1}{\mu_{i-1}} \right) = 0$,

distortion ,, $\Sigma \left\{ \Theta_i U_i^3 + \frac{1}{r_i} \left(\frac{1}{\mu_i} - \frac{1}{\mu_{i-1}} \right) U_i \right\} = 0$.

In each case it is assumed that the conditions occurring previously in the list are fulfilled.

It is however to be remembered that all these conditions have been derived on the supposition that terms of orders higher than the third in the angular aperture and angular field of view can be neglected: when the field of view is large, as in the case of photographic objectives, or when the angular aperture of the pencils is large, as in the case of microscope objectives, terms of higher order must be taken into account.

29. Herschel's condition.

Sir John Herschel formulated the condition which must be satisfied in order that an instrument, which is free from spherical aberration for the standard position of the object, may also be free from spherical aberration for positions of the object indefinitely near to this, *i.e.* that a slight displacement of the object along the axis may not introduce spherical aberration.

It was shewn by Abbe that this condition can be expressed in a form which is applicable to instruments of any aperture however large, just as the Fraunhofer condition for absence of coma can be extended in the form of the sine-condition. We shall first establish Abbe's condition, and then deduce Herschel's condition by supposing that the fifth power of the angular aperture can be neglected.

The condition in question, viz. that spherical aberration shall vanish for a second position of the object, adjacent to the one for which it is already known to vanish, is evidently equivalent to the condition that the magnification of a small segment of the axis, situated at the position of the object, may be the same whatever zone of the refracting surfaces is used to form the image. Let l be the length of this segment, l' the length of its image, μ and μ' the refractive indices of the initial and final media. Suppose that the image is formed by a thin meridian pencil whose chief ray makes an angle θ with the axis in the initial medium, and makes an angle θ' with the axis in the final medium. Applying Clausius' theorem (§ 7), we have

$$\mu l \sin\theta \,.\, d\theta = \mu' l' \sin\theta' \,.\, d\theta'.$$

Integrating this, $$\mu l \sin^2\frac{\theta}{2} = \mu' l' \sin^2\frac{\theta'}{2},$$

the constant of integration vanishing since θ and θ' vanish together.

Now the general equations of image-formation by paraxial pencils, namely (§ 8),

$$x' = -\frac{\mu'}{\mu}\frac{f^2}{x}, \qquad y' = \frac{fy}{x},$$

give $$\frac{dx'}{dx} = \frac{\mu'}{\mu} \,.\, \frac{f^2}{x^2} = \frac{\mu'}{\mu} \,.\, \frac{y'^2}{y^2},$$

so if m denotes the linear magnification of a small object at right angles to the axis, we have

$$\frac{l'}{l} = \frac{u'}{\mu} m^2.$$

Substituting in the preceding equation, we have

$$\mu'^2 m^2 \sin^2 \tfrac{1}{2}\theta' = \mu^2 \sin^2 \tfrac{1}{2}\theta$$

or

$$\frac{\sin \frac{1}{2}\theta'}{\sin \frac{1}{2}\theta} = \frac{\mu}{\mu' m}.$$

This is Abbe's condition: it is obviously impossible to satisfy it and the sine-condition simultaneously, save in exceptional cases.

We shall now proceed to derive Herschel's condition from this.

At the refraction at the ith surface, we have

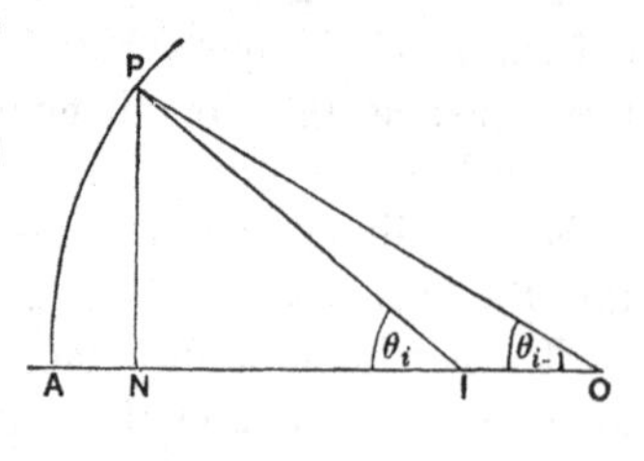

$$\sin^2 \tfrac{1}{2}\theta_{i-1} = \tfrac{1}{2}(1 - \cos\theta_{i-1}) = \frac{1}{2}\left(1 - \frac{NO}{PO}\right)$$

$$= \frac{1}{2}\left\{1 - \left(1 + \frac{PN^2}{NO^2}\right)^{-\frac{1}{2}}\right\}$$

$$= \frac{1}{4}\frac{PN^2}{NO^2} - \frac{3}{16}\frac{PN^4}{NO^4} \text{ approx.},$$

so

$$\sin \tfrac{1}{2}\theta_{i-1} = \frac{1}{2}\frac{PN}{NO}\left(1 - \frac{3}{8}\frac{PN^2}{NO^2}\right),$$

and

$$\frac{\sin \frac{1}{2}\theta_i}{\sin \frac{1}{2}\theta_{i-1}} = \frac{NO}{NI}\,\frac{1 - \frac{3}{8}\frac{PN^2}{NI^2}}{1 - \frac{3}{8}\frac{PN^2}{NO^2}},$$

or, in our usual notation,

$$\frac{\sin \frac{1}{2}\theta_i}{\sin \frac{1}{2}\theta_{i-1}} = \frac{s_i + \Delta_{i-1} - \frac{h_i^2}{2r_i}}{s_i' + \Delta_i - \frac{h_i^2}{2r_i}}\,\frac{1 - \frac{3}{8}\frac{h_i^2}{s_i'^2}}{1 - \frac{3}{8}\frac{h_i^2}{s_i^2}}.$$

If Abbe's condition is satisfied, we must therefore have

$$\prod_i \frac{\left(1 + \frac{\Delta_{i-1}}{s_i} - \frac{h_i^2}{2r_i s_i}\right)\left(1 - \frac{3}{8}\frac{h_i^2}{s_i'^2}\right)}{\left(1 + \frac{\Delta_i}{s_i'} - \frac{h_i^2}{2r_i s_i'}\right)\left(1 - \frac{3}{8}\frac{h_i^2}{s_i^2}\right)} = 1,$$

or

$$\sum_i \left(\frac{\Delta_{i-1}}{s_i} - \frac{\Delta_i}{s_i'} - \frac{h_i^2}{2r_i s_i} + \frac{h_i^2}{2r_i s_i'} - \frac{3}{8}\frac{h_i^2}{s_i'^2} + \frac{3}{8}\frac{h_i^2}{s_i^2}\right) = 0,$$

or

$$\sum_i \left\{\frac{\Delta_{i-1}}{s_i} - \frac{\Delta_i}{s_i'} + \frac{h_i^2}{2} Q_{si}\left(\frac{1}{\mu_i s_i'} - \frac{1}{\mu_{i-1} s_i}\right)\right\} - \frac{1}{8}\sum_i h_i^2\left(\frac{1}{s_i^2} - \frac{1}{s_i'^2}\right) = 0,$$

or

$$\sum_i \left\{\frac{\Delta_{i-1}}{s_i} - \frac{\Delta_i}{s_i'} + \frac{h_i^2 Q_{si}}{2}\left(\frac{1}{\mu_i s_i'} - \frac{1}{\mu_{i-1} s_i}\right)\right\} - \frac{1}{8}(\theta_0^2 - \theta_n^2) = 0,$$

since $h_i/s_i = \theta_{i-1}$.

The summation occurring here is the same as that occurring in the derivation of Fraunhofer's condition from the sine-condition: so the equation can at once be written in the form

$$\sum_i \left[\left\{ 1 + Q_{si} h_i^2 \sum_{p=1}^{i-1} \frac{d_p}{\mu_p h_p h_{p+1}} \right\} Q_{si} h_i^2 \left(\frac{1}{\mu_i s_i'} - \frac{1}{\mu_{i-1} s_i} \right) \right] - \frac{1}{4} (\theta_0^2 - \theta_n^2) = 0.$$

This is Herschel's condition. It is evidently compatible with the Fraunhofer condition only when $\theta_0 = \pm \theta_n$; this happens either when the object is at a point for which the angular magnification is ± 1, or when θ_0 and θ_n are both zero, *i.e.* when the system is telescopic and the object at infinity.

30. The impossibility of a perfect optical instrument.

Although it is possible to construct lens-systems satisfying the conditions which have been found, and therefore giving a satisfactory image for some definite position of the object when the aperture and field of view are not too large, we shall now shew that it is theoretically impossible to construct a really *perfect* optical instrument, *i.e.* one which will transform all points of the object-space into points of the image-space with some degree of magnification or minimisation. The proof is due to Klein*.

Suppose for the moment that such a perfect instrument exists. Since not only are points transformed into points, but lines (rays of light) are transformed into lines, the transformation of the object-space effected by the instrument is a collineation.

Now it is known that all the spheres of space have in common an imaginary circle at infinity, which contains the cyclic points of all the planes of the space†; an (imaginary) straight line which meets the circle at infinity is called a *minimal line*. Suppose then that the ray incident on one of the refracting surfaces of the instrument is a minimal line: the sine of the angle formed with the normal to the surface is infinitely great, and as conversely a minimal line is characterised by this infinitely large sine, it follows from the law of refraction that the refracted ray is also a minimal line.

This applies to each refraction; and therefore the collineation transforms each minimal line in the object-space into a minimal line in the image-space; so that the circle at infinity in the object-space is transformed into the circle at infinity in the image-space.

* *Zeitschrift für Math. u. Phys.* XLVI. (1901), 376.

† For two similar and similarly situated quadrics intersect in one plane curve at a finite distance and one at infinity: and spheres are similar quadrics.

From this it follows at once that the collineation is merely a similitude: it may be either direct or inverse (*i.e.* one which interchanges right and left).

In order to find the ratio of the similitude, suppose that c, c' denote the velocity of light in the object-space and image-space respectively. We can suppose that the similitude is direct, as if inverse it can be changed into a direct similitude by the addition of a plane mirror to the instrument.

Let the time taken by the light to travel from a point (x, y, z) to its image-point $(x', y' z')$ be denoted by $X(x, y, z)$. Let (x_1, y_1, z_1) be a point on one of the rays from (x, y, z) to (x', y', z'), at a distance r from (x, y, z); and let (x_2, y_2, z_2) be a point on another ray from (x, y, z) to (x', y', z'), also at a distance r from (x, y, z). Then if λ denote the ratio of similitude of the image-space and object-space, the distances of the image-points (x_1', y_1', z_1') and (x_2', y_2', z_2') from (x', y', z') are each λr, and (since the similitude is direct) they are each behind (*i.e.* beyond) (x', y', z'). The time from (x_1, y_1, z_1) to its image is therefore

$$X(x, y, z) - \frac{r}{c} + \frac{\lambda r}{c'};$$

and the time from (x_2, y_2, z_2) to its image is the same. So the times from (x_1, y_1, z_1) and (x_2, y_2, z_2) to their respective images are the same: but these are really arbitrary points in the object-space, so the time from any point in the object-space to its image is the same for all object-points. Hence we have

$$X(x, y, z) - \frac{r}{c} + \frac{\lambda r}{c'} = X(x, y, z)$$

or

$$\lambda = \frac{c'}{c},$$

so the dimensions of the object-space are to those of the image-space as c to c'. Thus when the instrument works in air, so that $c = c'$, the image is merely a life-size copy of either the object, or of the image obtained from the object by reflexion in a plane mirror.

31. Removal of the primary spectrum.

As already explained, the refractive index of a substance depends on the colour, *i.e.* the wave-length, of the light used in its determination. The behaviour of an optical system, which has been calculated in terms of the refractive indices, is therefore different for light of different colours: the position of the principal foci, the focal lengths, and the aberrations, will in general vary when the wave-length of the light is varied. As ordinary white light contains rays of all colours, there will therefore be a certain degree of confusion in the images formed by the optical instrument with white light: to this the name *chromatic aberration* is given. With a simple uncorrected lens of tolerably small aperture, the chromatic aberration is much more serious than the spherical aberration; with a convex lens of crown glass, if the red rays from a star are brought to a focus at a point R, the violet rays will intersect the plane through R perpendicular to the axis in a circle whose radius is about $\frac{1}{50}$ that of the lens, whatever be the focal length.

An optical system which is so contrived as to have the same behaviour for two standard wave-lengths is said to be *achromatic*. In order to achieve this, we must evidently secure that the row of images of the same object in light of different colours shall be doubled on itself, so that the images shall coincide in pairs: thus in an ordinary achromatic lens which is intended for visual observations, the yellow image is united with the dark green image, the orange-red with the blue, and the red with the indigo. Obviously at one end of this doubled row there must be two coincident images which differ infinitesimably in wave-length, *i.e.* there will be an image for which the rate of change of position with change of wave-length is zero: thus in the achromatic lens just mentioned, the images formed by the yellowish-green rays are closely united and focussed at minimum distance from the lens.

This pairing of images does not ensure an entire absence of chromatic aberration, since the images in three different colours will not coincide: but other terms, which will be mentioned later, are employed to denote a more complete freedom from colour troubles. The coloured fringes due to this outstanding colour-aberration are generally referred to as the *secondary spectrum*; a simple method (due to Sir G. Stokes) of observing the secondary spectrum of a lens is the following. Focus the lens on a vertical white line on a dark ground, and cover half the lens by a screen whose edge is vertical. Then evidently the yellow and green rays, which form an image nearer the lens than the mean image, will (coming from the uncovered half of the lens only) pass the mean image on one side of it, namely the side on which the screen is: while the red and blue rays, which form an image *beyond* the mean image, will pass on the other side of the mean image. The image will therefore have a citron-coloured margin on one side and a purple margin on the other.

32. Achromatism of the focal length.

The variation of behaviour of a transparent substance for light of different wave-lengths is usually measured by its *dispersion* or *dispersive power*,

$$\varpi = d\mu/(\mu - 1),$$

where μ is its refractive index for some standard wave-length and $\mu + d\mu$ is its index for some other standard wave-length not far removed from this.

Consider now the colour-variation of focal length of a single thin lens, for which we have (§ 11)

$$\frac{1}{f} = (\mu - 1)\left(\frac{1}{r} - \frac{1}{s}\right).$$

Differentiating this equation logarithmically, we have

$$\frac{df}{f} = -\varpi.$$

The focal length of a compound lens consisting of two thin lenses in contact, of focal lengths f_1 and f_2, is the reciprocal of

$$1/f_1 + 1/f_2:$$

so if the compound lens is to be achromatic, we must have

$$\frac{df_1}{f_1^2} + \frac{df_2}{f_2^2} = 0,$$

or

$$\frac{\varpi_1}{f_1} + \frac{\varpi_2}{f_2} = 0,$$

where ϖ_1 and ϖ_2 denote the dispersive powers. This equation represents the condition that the focal length, and consequently also in this case the position of the principal foci, may be the same for the two standard colours. *The combination is therefore achromatic for all distances of the object.*

The above equation shews that one of the lenses (say (1)) must be convergent and the other (say (2)) divergent: if the focal length of the whole is to be positive, we must have $f_1 < -f_2$, and consequently $\varpi_1 < \varpi_2$, so the divergent lens must have the greater dispersion. As flint glass has a greater dispersion than crown, the convergent lens is taken to be a crown and the divergent lens a flint. Roughly speaking, a flint whose diverging power is 2, will achromatise a crown whose converging power is 3, leaving a converging power of 1 for the compound lens.

The Petzval condition for flatness of field (§ 27),

$$\mu_1 f_1 + \mu_2 f_2 = 0,$$

requires however that μ_2 should be less than μ_1; so the convergent lens should have the higher refractive index, though having the smaller dispersive power, a condition which it was impossible to fulfil until the Jena glasses were introduced.

Consider next a system consisting of two thin lenses separated by an interval a. The focal length of the combination is (§ 11) the reciprocal of

$$\frac{1}{f_1} + \frac{1}{f_2} - \frac{a}{f_1 f_2},$$

so if the focal length is to be achromatised we must have

$$0 = -\frac{df_1}{f_1^2} - \frac{df_2}{f_2^2} + \frac{a\,df_1}{f_1^2 f_2} + \frac{a\,df_2}{f_1 f_2^2},$$

or

$$0 = \frac{\varpi_1}{f_1} + \frac{\varpi_2}{f_2} - \frac{a(\varpi_1 + \varpi_2)}{f_1 f_2}.$$

In such a system the two lenses would usually be of the same kind of glass, in order that whatever degree of achromatism is attained for two colours may as far as possible be attained for all colours: taking therefore ϖ_1 equal to ϖ_2, we have

$$0 = \frac{1}{f_1} + \frac{1}{f_2} - \frac{2a}{f_1 f_2},$$

or

$$a = \tfrac{1}{2}(f_1 + f_2),$$

so *the distance between the lenses must be half the sum of their focal lengths.* This condition is applied in the construction of eyepieces.

It is to be observed that the positions of the principal foci of the combination have not been achromatised, so that we have achromatised the size but not the position of the image. It is in fact impossible to achromatise a system of two non-achromatic lenses separated by a finite interval for *both* the size and position of the image: for if it were possible, the intermediate image, which is at the point where the line joining the object-point (supposed slightly off the axis) to the centre of the first lens intersects the line joining the image-point to the centre of the second lens, would be the same for every colour, and therefore each lens separately would be achromatic.

33. The higher chromatic corrections.

It is possible to remove the secondary spectrum, or more strictly speaking to replace it by a "tertiary" spectrum, by uniting the images in three instead of two colours. This can be effected for a combination of two lenses provided the Jena glasses are available: with the older glasses three lenses are required.

The variation of spherical aberration with the colour must also be taken into account. In the ordinary telescope objective, the citron image is corrected for spherical aberration, so the red image is under-corrected and the blue image is over-corrected: this defect is, in the case of the visual telescope, masked by the secondary spectrum: but with objectives of large angular aperture and short focal length, *e.g.* high-power microscope objectives, the correction of chromatic difference of spherical aberration is of greater importance than the elimination of the secondary spectrum.

Optical systems in which the spherical aberration is corrected for more than one colour, but in which the secondary spectrum is not removed, are called *semi-apochromatic*; while systems which have no secondary spectrum and are aplanatic (§ 24) for more than one colour are called *apochromatic*.

34. The resolving power of a telescope objective.

Nothing in our investigations hitherto has suggested the existence of any limit to the magnification attainable by means of an optical instrument; and it might therefore appear as if it were possible to construct a telescope of moderate dimensions which should reveal the minutest details of structure on the heavenly bodies. As a matter of fact, it is not possible, or at all events not profitable, to apply a magnifying power greater than a certain amount to a telescope with a given objective: and the reason for this is to be found in the circumstance that the wave-front by which the image of a star is formed is not a complete sphere, but is merely that fragment of a spherical wave which has been able to pass through the rim of the objective. This mutilated wave-front does not converge exactly to a point, as a full spherical wave would do, but forms a *diffraction pattern* in the focal plane of the objective, consisting of a bright disc whose centre is the image-point of the star as found by the preceding theory, surrounded by a number of dark and bright rings concentric with it

In order to determine the dimensions of this pattern, let a denote the diameter AB of the telescope objective, and S the centre of the diffraction pattern. The disturbance which is brought to a focus at a point T in the focal plane is the disturbance which at some preceding instant occupied the plane COD, perpendicular to the line OT which joins T to the centre O of the object-glass. Let $S\hat{O}T = \theta$, and let (ρ, ϕ) denote the polar coordinates of a point in the plane COD referred to O as origin and the line of greatest slope to the plane AOB as initial line.

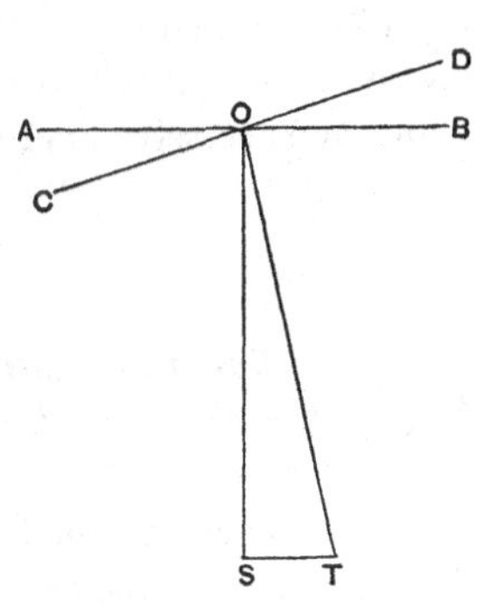

The disturbance at the point (ρ, ϕ) is proportional to

$$\rho d\rho d\phi \,.\, \sin 2\pi \left(\frac{t}{\tau} - \frac{z}{\lambda}\right),$$

where t denotes the time, τ the period of the light, λ its wave-length, and z the perpendicular distance of the point (ρ, ϕ) from the plane AOB.

But
$$z = \rho\theta \cos \phi,$$
θ being regarded as a small quantity.

The total disturbance at T is therefore

$$\iint \rho d\rho d\phi \,.\, \sin 2\pi \left(\frac{t}{\tau} - \frac{\rho\theta \cos \phi}{\lambda}\right),$$

integrated over the circle COD,

or
$$\sin \frac{2\pi t}{\tau} \int_0^{\frac{1}{2}a} \int_0^{2\pi} \rho d\rho d\phi \,.\, \cos \frac{2\pi\rho\theta \cos \phi}{\lambda},$$

since the elements of the integral involving $\sin \dfrac{2\pi\rho\theta \cos \phi}{\lambda}$ cancel each other in pairs.

Expanding the cosine in ascending powers of its argument, and integrating term by term, this becomes

$$\frac{\pi a^2}{4} \sin \frac{2\pi t}{\tau} \left\{1 - \frac{m^2}{2 \,.\, (1\,!)^2} + \frac{m^4}{3 \,.\, (2\,!)^2} - \frac{m^6}{4 \,.\, (3\,!)^2} + \dots\right\},$$

where m denotes $\pi\theta a/2\lambda$*.

* The series in brackets is a well-known Bessel-function expansion, being in fact $\frac{1}{m} J_1(2m)$. Cf. Whittaker, *Modern Analysis*, p. 267.

The first dark ring in the diffraction pattern will occur at the first point T for which the disturbance vanishes, *i.e.* it will correspond to the lowest value of m which makes the series in brackets to vanish: this is found by successive approximation to be

$$m = 1{\cdot}92,$$

giving

$$\theta = 1{\cdot}22\frac{\lambda}{a}.$$

The radius of the central diffraction-disc of a star (measured to the first dark ring) *formed in the focal plane of a telescope objective of aperture a and focal length f is therefore* $1{\cdot}22\lambda f/a$.

A telescope is usually estimated to succeed in dividing a close double star when the centre of the diffraction-pattern of one star falls on the first dark ring of the diffraction-pattern of the other star: when this is the case, it follows from the preceding equation that the angular distance between the stars in seconds of arc is $1{\cdot}22 \times 206265 \times \lambda/a$. If we express a in inches, and take $\lambda = 1/50{,}000$, this gives for the angular distance between the stars,

$$\frac{5''}{\text{aperture in inches}}.$$

This is known as *Dawes' rule for the resolving power of a telescope objective.*

35. The resolving power of spectroscopes.

The power of spectroscopic apparatus (prisms or gratings) to separate close spectral lines involves the same principles as the power of telescope objectives to separate the components of double stars. Each spectral line is really a diffraction-pattern, consisting of a narrow bright band at the place of the geometrical image of the line, flanked by alternate bright and dark bands: and the spectroscope is said to *resolve* two lines of adjacent wave-lengths when the centre of the central bright band arising from one wave-length falls on the first dark band of the pattern arising from the other wave-length.

The difference between the telescopic and the spectroscopic cases is that in the telescope we are dealing with the circular diffraction-pattern of a point-source, formed by a circular beam, while in the spectroscope we are dealing with the banded pattern of a line-source, formed by a beam of rectangular cross-section. The latter case is analytically the simpler of the two, since all sections of the beam at right angles to the spectral lines are similar to each other, and the problem can therefore be treated as a two-dimensional one.

Let AB be the wave-front, limited by an aperture AB of breadth a, of a pencil of parallel light of wave-length λ, representing one of two vibrations which are just resolved. Draw BC at right angles to AB, and take BC equal to λ. Then the first dark band of the diffraction-pattern corresponding to the disturbance AB will fall at the place to which a wave-front occupying the position AC is brought to focus: for the phase of the AB-disturbance at C differs by a whole wave-length from the phase at B, *i.e.* from the phase at A, and consequently every point in AC will have a corresponding point in the other half of AC which is in exactly the opposite phase, and so will interfere with it to produce total darkness.

Thus the disturbances represented by AB and AC will be just resolved: so *if* $\delta\theta$ *denote the angle between two wave-fronts of approximate wave-length* λ, *the disturbances will be just resolvable when the beams are of breadth* a, *provided that*

$$\delta\theta = \frac{\lambda}{a}.$$

As the product of the inclination of two plane wave-fronts and their diameter is, by Helmholtz's theorem, unaltered by passage through any system of lenses, it is evident that the resolvability of two adjacent disturbances is not altered by passage through any lens-system which does not introduce new diaphragm limitations, and so depends solely on the prisms or grating.

The *resolving power* of a spectroscope is defined by Lord Rayleigh to be $\lambda/\delta\lambda$, when two spectral lines of wave-lengths λ and $\lambda + \delta\lambda$ respectively can just be resolved, the slit of the spectroscope being infinitely narrow. But the result obtained above gives

$$\frac{\lambda}{\delta\lambda} = a\frac{d\theta}{d\lambda}.$$

Thus *the resolving power of any grating or train of prisms is measured by the product of the breadth of the emergent beam of parallel light and the dispersion*; the dispersion being defined as the rate of change of deviation with wave-length*.

* The resolving power can be regarded from a different point of view as equal to the number of separate pulses into which a single incident pulse of light is broken up by the spectroscope. For references to this aspect of the theory, cf. a paper by the author in *Monthly Notices of the Royal Astron. Society*, LXVII. p. 88.

CHAPTER III.

SKETCH OF THE CHIEF OPTICAL INSTRUMENTS.

36. The photographic objective.

The simplest form of photographic objective is a single convergent lens*; the light from an object at some distance is rendered convergent by the lens, and the real image thus formed is received on a gelatine film containing emulsified bromide of silver: this salt is acted on by light, and after undergoing the processes of development and fixation yields a permanent image in metallic silver.

The *rapidity of action* of the lens depends only on its *aperture-ratio*, which is the ratio of its focal length f to its diameter: if the diameter be f/n, the time of exposure required is proportional to n^2; for the exposure is inversely proportional to the light falling on unit area of the image, and is therefore proportional to the area of the image divided by the total light received by the lens from the object: but the area of the image is proportional to f^2, and the total light received is proportional to the area of the lens, *i.e.* to $(f/n)^2$: so the time of exposure is proportional to n^2.

This theorem applies equally to objectives which are not constituted of a single lens, provided that instead of the diameter of the lens we take the diameter of the entrance-pupil. For an average photographic objective, n is about 7 when the full aperture is used: for portrait lenses, which are very rapid, n may be as low as 3.

* If a single non-achromatic convergent lens were used, it would be best to select a deep meniscus with its concavity turned towards the object (this secures considerable depth of focus and a large field of fair definition) and to use a narrow stop in front (which reduces the spherical aberration and curvature of image): when the focus is being obtained, a weak convex lens must be inserted, so as to reduce the visual focal plane to the place which the focal plane of the actinic rays occupies when this lens is absent.

The single convergent lens is practically useless, on account of the defects which have been discussed in the preceding chapters; and it is necessary to design objectives formed of more than one lens, with a view to the special requirements of terrestrial photography, which are the following:

1. The system must be achromatised in such a way that the visual image, which is used in finding the focus, may coincide with the actinic image which acts on the sensitive plate: the D line of radiation of sodium is generally united with the blue H_β radiation.

2. The definition should be such that points of the object are represented by dots of (say) not more than $\frac{1}{100}$ of an inch diameter, over a field of (say) 50° square: in the case of portrait lenses, this requirement is sacrificed in order to obtain the greatest possible rapidity: a portrait lens will not usually cover a greater field than about 25° square. In any case, the standard of definition is much lower than is demanded of telescope objectives, but the field is much wider. The definition is usually improved by *stopping down*, *i.e.* narrowing the aperture of the diaphragm: but this involves a loss of rapidity.

3. Distortion must as far as possible be eliminated: objectives consisting of lenses cemented together, with the stop in front, always give barrel distortion (§ 28), while if the stop is between the lens and the image there is pincushion distortion. If we combine these two systems into a *doublet*, *i.e.* a system of two compound lenses separated by an interval in which the stop is placed, the two opposite distortions neutralise each other and we obtain an objective which is *rectilinear*, *i.e.* free from distortion.

4. The objective must have a certain amount of *depth of focus*, *i.e.* must be able to give fairly sharp images of objects which are in front of or behind that object-plane which is accurately focussed. Depth of focus is usually measured by the range of object-distance for which the pencil meets the sensitive plate in a disc of less than a certain diameter: it depends on the object-distance, focal length, and aperture, but does not vary much with the type of lens. The depth of focus is obviously increased by stopping down, since then all pencils become narrower: with equal aperture-ratios, the depth is greater for small focal lengths than for large ones.

5. Among minor requirements may be mentioned freedom from *flare*, *i.e.* from light which has been reflected at some of the refracting surfaces, and which on reaching the sensitive plate interferes with the brilliancy of the image.

It is of course not possible here to enter into details regarding the construction of the various types of photographic objective which are at present on the market.

37. Telephotography.

When an object at a great distance is photographed with an ordinary photographic objective, the image is inconveniently small and the details difficult to distinguish. A more convenient image can be obtained by making an enlargement from this photograph: but owing to the grain of the sensitive plate, and the insufficient definition of the primary image, it is not practicable to enlarge many diameters. It is therefore desirable to obtain a primary image as large as possible. Now in order to obtain a large-scale image, the camera must have an objective of great focal length: and as with most objectives the length of the camera is nearly equal to the focal length, this requires an inconvenient or impossible extension of the camera. The difficulty is surmounted by removing the principal point of the system (which is at a distance from the principal focus equal to the focal length, and is generally near the objective) to a considerable distance in front of the objective, so that although the focal length is great, the distance from the objective to the sensitive plate is comparatively small. This is effected in *telephotography*, in which a divergent lens is introduced between the convergent objective and the sensitive plate: this divergent lens diminishes the convergence of the pencils which fall on it from the convergent combination, so that they become practically the same as the pencils which would have proceeded from a convergent lens of great focal length, placed at a considerable distance in front of the actual position of the objective.

38. The telescope objective.

The conditions which must be satisfied by the objectives used in astronomical telescopes, whether visual or photographic, differ greatly from the conditions which must be satisfied by the objectives used in terrestrial photography. In the latter, definition which will bear a feeble magnification is required over a field of (say) 50° square: in the former, definition which will bear a much higher magnification is required, but over a much smaller field: the field seen at one time in a large visual telescope is only about $\frac{1}{4}°$ in diameter, and the region depicted on the sensitive plate of a photographic telescope is usually only of about the order of magnitude of a square degree. Consequently the defects of astigmatism, curvature of field, and distortion, which come into prominence at the outer parts of a wide field, are much less important in celestial than in terrestrial work: while on the other hand the defects of spherical aberration and coma, which affect the central parts of the field, must be more carefully eliminated in the astronomical objective than in the ordinary photographic lens. Moreover, since any diminution in light-gathering power is to be avoided at all costs in astronomy, it is not permissible to correct errors by means of diaphragm effects. For these reasons the doublet, which is predominant in terrestrial photography, is abandoned by astronomers in favour of an objective consisting of two or three lenses fairly close together, designed to make the corrections for spherical aberration and coma as perfect as possible.

The colour corrections also differ in the two cases. In the terrestrial lens, the actinic image must be made to coincide in position with the visual image which is used in focussing: but as in astro-photographic work the focus is found by taking trial plates, there is no need to trouble about the visual rays, and consequently the colour correction can be devoted wholly to the improvement of the actinic image, the blue H_β radiation being generally united with a violet radiation emitted by mercury. In the visual telescope there is no need to take account of the actinic image, and the yellowish-green rays are brought to the minimum focus.

We shall now shew how the equations found in Chapter II can be applied to design what may be called a *Fraunhofer objective*: this will be defined as a telescope objective consisting of two lenses whose thickness will be neglected, in contact at their vertices, and having their four radii of curvature chosen to satisfy the following conditions

(i) Given focal length F for the objective as a whole,

(ii) Achromatism,

(iii) Absence of spherical aberration for an object at infinity, neglecting the 5th power of the aperture,

(iv) The sine-condition for an object at infinity, neglecting the 5th power of the aperture.

Let r_1, r_2, r_3, r_4 denote the radii of the refracting surfaces in order (all taken positively when convex to the incident light), μ, μ' the refractive indices of the lenses, ϖ_1, ϖ_2 their dispersions for the radiation which it is desired to have at minimum focus, f_1 and f_2 their focal lengths.

Conditions (i) and (ii) may be written

$$\frac{1}{F}=\frac{1}{f_1}+\frac{1}{f_2},\qquad \frac{\varpi_1}{f_1}+\frac{\varpi_2}{f_2}=0.$$

Thus if κ denote ϖ_1/ϖ_2, we have

$$\frac{1}{f_1}=\frac{1}{F(1-\kappa)},\qquad \frac{1}{f_2}=-\frac{\kappa}{F(1-\kappa)},$$

and therefore

$$\frac{1}{r_1}-\frac{1}{r_2}=\frac{1}{F(1-\kappa)(\mu-1)}\ \text{ and }\ \frac{1}{r_3}-\frac{1}{r_4}=-\frac{\kappa}{F(1-\kappa)(\mu'-1)}.$$

These equations are satisfied identically if we write

$$r_1=\frac{F(1-\kappa)}{p_1},\qquad r_2=\frac{F(1-\kappa)}{p_1-\dfrac{1}{\mu-1}},\qquad r_3=\frac{F(1-\kappa)}{p_2+1},\qquad r_4=\frac{F(1-\kappa)}{p_2+1+\dfrac{\kappa}{\mu'-1}},$$

where p_1 and p_2 are now to be determined from conditions (iii) and (iv).

These conditions (iii) and (iv) are

(Spherical aberration condition, § 20)

$$Q_1^2u_1+Q_2^2u_2+Q_3^2u_3+Q_4^2u_4=0,$$

(Sine-condition, § 25)

$$Q_1u_1+Q_2u_2+Q_3u_3+Q_4u_4=0,$$

where

$$Q_1=\frac{1}{r_1}-\frac{1}{x_1}=\mu\left(\frac{1}{r_1}-\frac{1}{x_2}\right),$$

$$Q_2=\mu\left(\frac{1}{r_2}-\frac{1}{x_2}\right)=\frac{1}{r_2}-\frac{1}{x_3},$$

and similar equations hold for Q_3 and Q_4: x_1, x_2, x_3, x_4, x_5, being the distances of the object and its successive images from the objective; and where

$$u_1=\frac{1}{\mu x_2}-\frac{1}{x_1},\qquad u_2=\frac{1}{x_3}-\frac{1}{\mu x_2},\qquad u_3=\frac{1}{\mu' x_4}-\frac{1}{x_3},\qquad u_4=\frac{1}{x_5}-\frac{1}{\mu' x_4}.$$

Now since the object is at infinity, we have

$$\frac{1}{x_1}=0,\qquad \frac{1}{x_2}=\frac{\mu-1}{\mu r_1},\qquad \frac{1}{x_3}=(\mu-1)\left(\frac{1}{r_1}-\frac{1}{r_2}\right),$$

$$\frac{1}{x_4}=\frac{\mu-1}{\mu'}\left(\frac{1}{r_1}-\frac{1}{r_2}\right)+\frac{\mu'-1}{\mu' r_3},\qquad \frac{1}{x_5}=\frac{1}{F},$$

and consequently

$$Q_1=\frac{1}{r_1},\qquad Q_2=\frac{\mu}{r_2}-\frac{\mu-1}{r_1},\qquad Q_3=\frac{1}{r_3}-(\mu-1)\left(\frac{1}{r_1}-\frac{1}{r_2}\right),\qquad Q_4=\frac{1}{r_4}-\frac{1}{F}.$$

Substituting in terms of p_1 and p_2, we have (neglecting common factors)

$$Q_1 = p_1, \quad Q_2 = p_1 - \frac{\mu}{\mu - 1}, \quad Q_3 = p_2, \quad Q_4 = p_2 + \frac{\mu'\kappa}{\mu' - 1},$$

$$u_1 = \frac{(\mu - 1)p_1}{\mu^2}, \quad u_2 = 1 - u_1, \quad u_3 = \frac{\mu' - 1}{\mu'^2} p_2 - \frac{\mu' - 1}{\mu'}, \quad u_4 = -\kappa - u_3.$$

Substituting in the conditions (iii) and (iv) above, we have

$$\frac{p_1}{\mu}\left(2p_1 - \frac{\mu}{\mu - 1}\right) + \left(p_1 - \frac{\mu}{\mu - 1}\right)^2$$
$$- \kappa\left(-1 + \frac{p_2}{\mu'}\right)\left(2p_2 + \frac{\mu'\kappa}{\mu' - 1}\right) - \kappa\left(p_2 + \frac{\mu'\kappa}{\mu' - 1}\right)^2 = 0,$$

and
$$\frac{\mu + 1}{\mu} p_1 - \kappa \frac{\mu' + 1}{\mu'} p_2 - \frac{\mu}{\mu - 1} + \kappa - \frac{\mu'\kappa^2}{\mu' - 1} = 0.$$

The second equation gives p_2 as a linear function of p_1; substituting in the preceding equation, we have a quadratic for p_1, which can be solved: the radii of curvature of the surfaces of the objective are thus determined.

39. Magnifying glasses and eyepieces.

For the rough examination of small objects, the *magnifying glass* is used. This in its simplest form consists of a single convergent lens, held between the eye and the object, at a distance from the latter somewhat less than its own focal length: an enlarged virtual image is thus formed at some distance behind the object, and this is examined by the eye. The pupil of the eye is the diaphragm effective in limiting the aperture of the image-forming pencils, and the rim of the lens (supposing it to be of greater diameter than the pupil) is the diaphragm effective in limiting the field of view.

Closely allied to the magnifying glasses are the *eyepieces* which are used to examine the images formed by the objectives of visual telescopes and microscopes. These consist usually of two lenses separated by an interval: the lens which is nearest the eye is called the *eye-lens*, and the other the *field-lens*.

In *Huyghens' eyepiece* the lenses are placed at a distance apart equal to half the sum of their focal lengths, in order to satisfy the condition of achromatism found in § 32. The focal length of the field-lens is usually three times that of the eye-lens, but in some modern eyepieces, especially those used for low-power magnification with the microscope, the ratio of the focal lengths is smaller than this. The lenses used are plano-convex, with the convex sides turned towards the image to be examined.

The first principal focus of Huyghens' eyepiece falls between the lenses, and consequently the image to be examined (which must be placed at this point in order that the emergent wave-fronts may be plane) can only be a virtual image: in other words, a Huyghens' eyepiece, when used with a telescope objective, must be pushed in nearer to the objective than the place at which the objective would form a real image of the object. On this account the Huyghens construction cannot be used in micrometer eyepieces, in which it is desired to place a framework of spider-lines in the plane of the image formed by the objective, and to examine the spider-lines and the image together by the eyepiece.

The image formed by high-power apochromatic microscope objectives is usually examined by a *compensating eyepiece*, which is specially corrected chromatically in order to neutralise the chromatic difference of magnification due to the objective.

In *Ramsden's* construction, which is always used in micrometer eyepieces, the first focal plane of the combination does not fall between the lenses, and the eyepiece can consequently be used in order to simultaneously examine the image (formed by the objective of the telescope or microscope) and also a reticle of spider-lines, placed in its plane with a view to micrometric measurements. In this construction the two lenses are usually plano-convex with the convex sides turned towards each other: they are taken to be of the same focal length, and therefore if the condition of achromatism were satisfied the interval between the lenses would be exactly equal to this focal length: with this arrangement however the field lens would be exactly in the focus of the eye-lens, which is undesirable; and the interval is consequently taken to be shorter than the focal length, the resulting chromatic error being (in the best eyepieces) removed by substituting achromatic combinations for the simple field-lens and eye-lens.

The field-lens is so near to the real image examined, that its principal effect is to deflect the chief rays of the pencils towards the axis of the instrument, without greatly altering the inclination of the other rays of the pencils to the chief rays: the function of magnifying is therefore performed almost wholly by the eye-lens.

40. The visual astronomical refractor.

The astronomical refracting telescope, as used visually, is formed by the combination of an astronomical objective (§ 38) with an eyepiece (§ 39) which is used to examine the image formed by it. In the typical normal case the eyepiece is so placed that its first focal plane coincides with the second focal plane of the objective : under these circumstances the parallel pencil of light from a star is made to converge to an image situated in this focal plane, and is re-converted into a parallel pencil by the eyepiece. Short-sighted observers find it convenient to push the eyepiece nearer to the objective, so that the emergent pencils are divergent.

The diaphragm effective in limiting the apertures of the pencils is the rim of the objective: this is therefore the entrance-pupil (§ 16). The exit-pupil, which is the image of the objective formed by the eyepiece, is outside the instrument and behind it, and the eye is placed there. The field of view is generally limited by a diaphragm placed in the focal plane of the objective : if this were not present, the field would be limited by the rim of one of the lenses of the eyepiece, and there would be a "ragged edge" of the field seen only by partial pencils. The field of view is of course the angle subtended at the centre of the objective by this diaphragm.

The magnifying power (§ 17) is readily seen to be the ratio of the focal lengths of the objective and eyepiece. A telescope is usually furnished with a battery of eyepieces, giving various magnifications.

When the eyepiece is of such short focal length that the magnifying power of the telescope is greater than a number which may be roundly stated as equal to the diameter of the objective in millimetres, the definition is spoiled by the diffraction effects discussed in § 34: from this to one-half of it may be regarded as the useful range of magnifying power, since below this limit the capabilities of the objective are not being used to their full extent. This corresponds to an exit-pupil of 1 to 2 mm., which is much smaller than the pupil of the eye.

If the object viewed is a star, which may be regarded as a mathematical point, the brilliancy varies directly as the light gathered by the objective, *i.e.* as the square of the aperture, and is independent of the focal length. The same consideration applies to the rapidity of an astro-photographic objective.

The aperture-ratio (§ 36) of a telescope objective is usually about 15: but for small telescopes it is frequently smaller. In the old telescopes, which were constructed before the discovery of achromatic combinations, the aperture-ratio was very large: this was in order to take advantage of the fact that the influence of chromatic aberration on the distinctness of an object is inversely proportional to the aperture-ratio.

41. The astronomical reflector.

In the astronomical reflecting telescope, the light from a celestial object is received on a concave mirror, which serves the same purpose as the objective of a refracting telescope, namely to form a real image of the object in its own focal plane. This image can either be allowed to impress itself directly on a sensitive plate, or may be examined by an eyepiece. In the latter case, it is necessary to insert a small plane

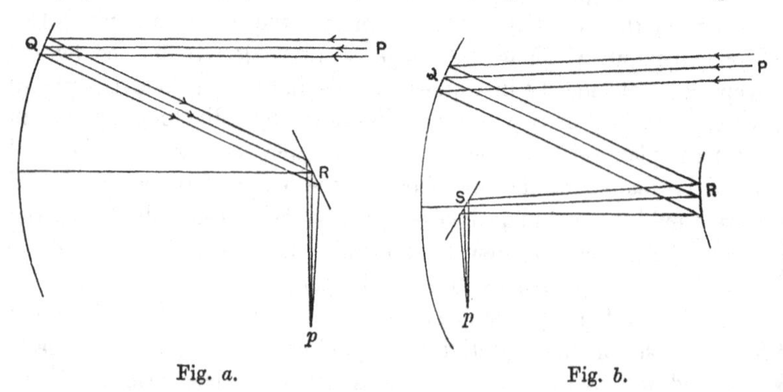

Fig. *a*. Fig. *b*.

mirror obliquely in the path of the rays after leaving the large mirror, in order to divert them to the side of the telescope, where the image is formed and examined: otherwise the head of the observer would obstruct the passage of the incident light to the large mirror. This construction is known as the *Newtonian reflector* (Fig. *a*): the path of the rays from a star P to its real image p will be obvious from the diagram, Q being the large mirror and R the flat. The magnifying power, as in the case of the refractor, is the ratio of the focal lengths of the objective and eyepiece.

In certain cases, *e.g.* the photography of planets, it is desirable to obtain on the sensitive plate an image on a larger scale than would be furnished directly by the concave mirror: this is achieved by making use of what is essentially the same principle as that on which telephotography (§ 37) is based, namely receiving the rays from the large mirror on a small divergent (*i.e.* convex) mirror before allowing them to form a real image. This is known as *Cassegrain's* construction (Fig. *b*). The path of the rays from the star P to its real image p, after reflexion at the large mirror Q, the convex mirror R, and the flat S, will be obvious from the diagram.

The diaphragm effective in limiting the aperture of the image-forming pencils of a reflector is the rim of the large mirror. The field of view of a visual reflector is limited by the rim of one of the eyepiece lenses, or by a diaphragm placed in the plane of the real image in order to exclude the part of the image formed by partial pencils.

The correction for spherical aberration of the large mirror is effected by figuring it to a paraboloidal form: as we have seen however (§ 24) this does not remove coma, which is accordingly an outstanding defect in all reflecting telescopes.

The reflector is of course perfectly free from chromatic aberration, and this involves a further advantage over the refractor in that it permits the construction of reflectors having a much smaller aperture-ratio than refractors, and consequently much greater rapidity for objects with an extended area.

The aperture-ratio of the large mirror of a modern reflector is usually about 5: the addition of a convex mirror, which usually gives about a threefold magnification, raises the aperture-ratio to about 15 in Cassegrain's construction.

For the above reasons, and also because it is easier to construct a mirror than an objective of the same diameter, and therefore easier to secure light-gathering power, the reflector is specially suited for the photography of nebulae.

42. Field, Marine, and Opera Glasses.

The visual astronomical telescope cannot be applied to terrestrial uses without modification, since the image which is formed by the objective and examined by the eyepiece is inverted. It is possible to surmount this difficulty by the use of an *erecting eyepiece*, which is in principle similar to the microscope (§ 43), forming a second (erect) real image in its interior: but the instrument so constituted is of considerable length and cannot be supported steadily in the hands without difficulty. Accordingly field-glasses were until recent years always formed of a convergent objective combined with a divergent

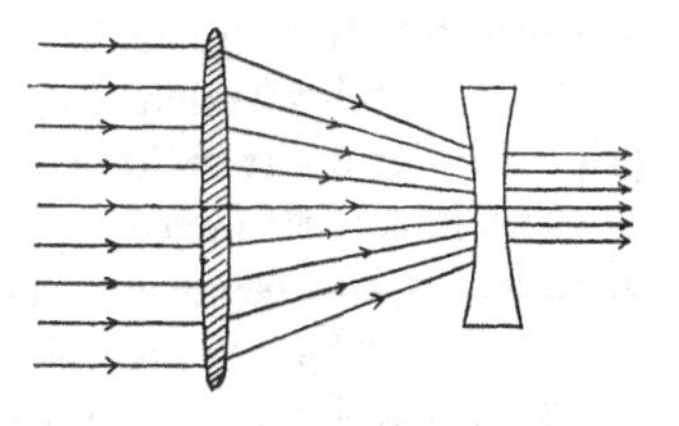

eyepiece: the rays after leaving the objective and before reaching the plane of the real image were intercepted by the eyepiece, which destroyed their convergence and rendered them parallel at emergence. The path of the rays will be evident from the diagram.

Since no real image is formed in this construction, which is known as the *Galilean telescope*, there is no inversion of the object. The diaphragm effective in limiting the field of view is the rim of the object-glass, and the diaphragm effective in limiting the aperture of the pencils is the pupil of the eye. The magnifying power, as in other telescopes, is the ratio of the focal lengths of the objective and eyepiece.

The Galilean telescope has a much smaller field of view than an astronomical telescope of the same magnifying power; on this account the best modern field-glasses have reverted to the astronomical type of telescope, with a device suggested originally by Porro for re-erecting the object and shortening the tube-length of the telescope. This device, which is represented in the annexed diagram, is to interpose a prism in the path of the light when it has travelled some distance from the objective: the rays fall normally on the hypotenuse face of the prism, and after passing through the glass to one of the other faces are totally reflected, passing thence to the third face where they are again totally reflected: after this they travel through the glass to the hypotenuse face again and emerge normally from the prism. The effect of the two total reflexions has been to reverse the direction of the beam, so that the rays are now travelling back towards the objective: after proceeding some distance in this direction they are again intercepted by a double-total-reflexion prism, with its principal section at right angles to that of the first prism: this once more reverses the direction of the beam and sends it on to the eyepiece, whence it passes into the eye. A field-glass, formed of two telescopes of

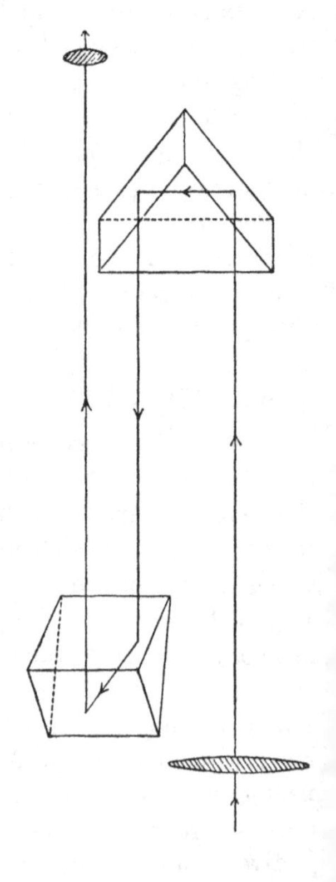

this construction (one for each eye) is called a *Prismatic Binocular*: the folding up of the path of the rays by the two reversals greatly reduces the length of the instrument, and the total reflexions perform the other necessary function of erecting the image. The magnifying power of a Prism Binocular usually ranges from 6 to 12, and the field ranges from 3° to 8° in diameter.

43. The Microscope.

The simple magnifying glass (§ 39) cannot advantageously be constructed to give magnification above a certain limit, owing in part to the excessive smallness of the lens which would be required for a high magnification. In order to pass beyond this limit, we can conceive an astronomical telescope placed immediately behind the magnifying glass, so that the pencil from a point of the object off the axis, after being converted by the magnifying glass into a pencil of nearly parallel rays, passes through the telescope and thereby increases its angle of divergence from the axis of the instrument. In this way we attain a magnifying power which is roughly the product of the magnifying powers of the magnifying glass and the telescope.

This arrangement is essentially a *microscope*, the combination of the magnifying glass and telescope objective being called the *objective* of the microscope, and the telescope eyepiece being the *eyepiece* of the microscope. The object to be viewed is placed in front of the microscope objective (which always consists of a combination of several lenses, and has a very short focal length) at a distance from it slightly greater than the focal length: a real enlarged image is consequently formed by the objective and examined by an eyepiece.

The *magnifying power* of the entire instrument, which we have defined in § 17 as the ratio of the linear dimensions of image and object when the image is at the standard distance of distinct vision, is readily found to be approximately equal to

$$\frac{\text{Length of tube} \times \text{Conventional distance of distinct vision}}{\text{Focal length of objective} \times \text{Focal length of eyepiece}}.$$

A microscope objective must be designed to give the best possible definition when a small field of view is seen by pencils of very wide angular aperture: the incident cones of light have apertures as great as 150°. Consequently of the aberrations discussed in Chapter II, the most important in the construction of microscope objectives are, spherical aberration, coma (the sine-condition), and chromatic aberration.

The pencils are of such wide angle that spherical aberration must be much more completely removed than would be the case by the satisfaction of the approximate condition found in § 20; this further spherical correction is usually known as "spherical zones." Moreover the same circumstance—the wide angle of the pencils—causes the chromatic variation of the spherical aberration (§ 33) to assume serious proportions, and in all good objectives it is specially corrected. In the best or *apochromatic* objectives (§ 33), the secondary spectrum is also removed.

In high-power objectives, advantage is taken of the property of the aplanatic points of the sphere discussed in § 24; the front lens of the objective is a hemisphere with its plane face turned towards the object: below this is a film of cedar-wood oil (whose refractive index, 1·51, is practically the same as that of the hemisphere), separating the objective from a cover-glass, usually 0·18 mm. thick, which protects the object. In this way the object is virtually within a sphere whose refractive index is that of the glass, and in fact is situated at the internal aplanatic point of the sphere, a magnified image being formed at the external aplanatic point.

An immersion objective (*i.e.* one in which the oil is used) collects a wider cone of light from the object than a dry objective would do: for if the cone of light on emerging from the cover-glass passes into air (as happens with dry objectives), its rays are bent outwards by the refraction, and consequently the outermost rays of the cone will pass outside the rim of the objective; in the immersion objective they are not refracted on emergence from the cover-glass, and so pass on into the objective.

We must now discuss the resolving power of the microscope. The object will first be treated as if it were self-luminous, ignoring the fact that it is actually seen by light directed on it from another source.

Let the semi-vertical angle of the cone of light issuing from the object to the objective be θ, and let the semi-vertical angle of the cone forming the image be θ': let μ denote the refractive index of the cedar-wood oil, μ being replaced by unity in the case of dry objectives. The quantity $\mu \sin \theta$ is called the *numerical aperture* of the objective, and is generally denoted by the letters N.A.*

The wave-front from the object, being limited by the rim of the

* It is approximately equal to the ratio of the radius of the back lens of the objective to the focal length of the objective.

objective, will form a diffraction-pattern at the image; regarding the objective as compounded of a magnifying-glass and a telescope-objective in juxtaposition, we can apply the theorem of § 34, which at once shews that the radius of the central diffraction-disc at the image is

$$\frac{1 \cdot 22\,\lambda}{2 \tan \theta'},$$

where λ is the wave-length of the light. If m denote the magnification, it follows that the centre of the image of one object will fall exactly on the first dark ring of the diffraction-pattern of a second object, provided the distance apart of the objects is

$$\frac{0 \cdot 61\,\lambda}{m \tan \theta'}.$$

Now the sine-condition gives the equation

$$\frac{\mu \sin \theta}{\sin \theta'} = m, \quad \text{or} \quad N.A. = m \sin \theta',$$

and as $\sin \theta'$ and $\tan \theta'$ are practically equal (θ' being a small angle), we see that *the distance apart of two objects which can just be resolved is*

$$\frac{0 \cdot 61\,\lambda}{N.A.}.$$

The best immersion objectives have a numerical aperture of 1·4: taking $\lambda = \cdot 0005$ mm., we see that two objects which can just be resolved with these objectives will be approximately at a distance apart equal to

$$\frac{0 \cdot 61 \times \cdot 0005}{1 \cdot 4} \text{ mm., or } \cdot 00022 \text{ mm.}$$

In this discussion we have however neglected one fact of importance, namely that the object studied by the microscope is not truly self-luminous, but is illuminated by another source of light. The importance of this distinction was first shewn by Abbe, who observed that the light incident from the source is diffracted by the object, and that in order to obtain an image correctly representing the structure of the object it is essential that the objective should receive the whole of this diffraction-pattern. If this condition is not satisfied, the image obtained will represent a fictitious object, such as would give rise to a diffraction-pattern consisting of those parts of the actual diffraction-pattern which are transmitted by the objective.

44. The Prism Spectroscope.

A *spectroscope* is an instrument designed for the work of analysing any given composite radiation into its constituent simple radiations, each with its own wave-length. In the *prism spectroscope*, this is done by taking advantage of the fact that the refractive index of glass for any kind of light depends on the wave-length of the light, and that consequently radiations of different wave-lengths can be separated from each other by causing them to pass through a glass *prism*, *i.e.* a piece of glass bounded by two optically-plane faces inclined to each other.

If for example the light which it is desired to analyse is that produced by the flame of a Bunsen burner, in which a salt of sodium is volatilised, the usual practice is to throw an image of the flame (by means of a convergent lens) on a narrow slit between two jaws of metal, so that the opening of the slit is strongly illuminated by the yellow light. This slit is placed in the focal plane of a telescope objective, so that the sodium light which is able to pass between the jaws of the slit travels on to the objective and is there converted into a parallel beam. In this condition it is received on one face of a prism, and passes through the glass and out at the other face; the beam is then received normally on another telescope objective, in the focal plane of which two images of the slit are formed close together; these images correspond to two kinds of yellow radiation emitted by the sodium flame, which have followed slightly different paths in the prism and have thus become separated. Each kind of radiation emitted by the original source of light will give rise in this way to a distinct final image of the slit: these slit-images are called *spectral lines*, and collectively form the *spectrum* of the source of light: they may be allowed to impress themselves on a sensitive plate, or may be examined visually with an eyepiece.

The slit and the first telescope objective are together called the *collimator*: and the collimator, prisms (the light may pass through more than one prism successively), and final telescope, constitute a *prism spectroscope*.

We shall first find the dispersion produced by the train of prisms, *i.e.* the differential effect of the prisms on two radiations of slightly different wave-lengths. Suppose that the light consists of two kinds of radiation, for one of which the refractive index is typified by μ and for the other by $\mu + \delta\mu$: and let $\delta\theta$ denote the angle between the two emergent beams corresponding to these two kinds of radiation: we shall now find $\delta\theta$*.

* The method is due to Lord Rayleigh.

Let PQ be a wave-front at incidence on the prism-train, $P'Q'$ the corresponding piece of a wave-front for the radiation μ after emergence from the prisms into air. PP' and QQ' the paths of the rays from P to P' and from Q to Q' for this radiation, RQ' and SP' the paths from the wave-front PQ to Q' and P' for the light $\mu + \delta\mu$, and T the point in which the path SP' meets the wave-front of the light $\mu + \delta\mu$ through Q'.

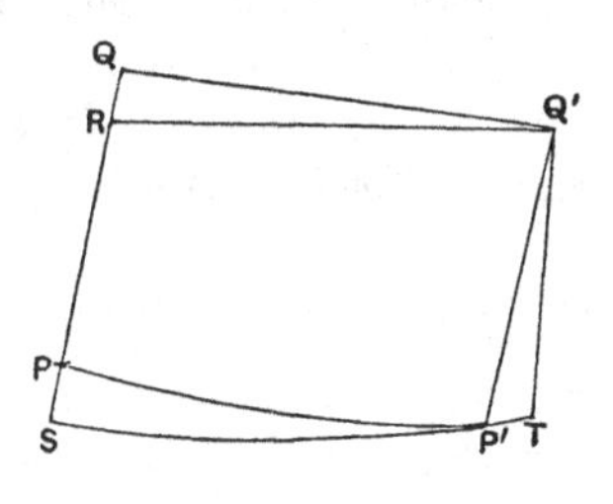

Then $\delta\theta = P'\hat{Q}'T = \dfrac{P'T}{P'Q'}$

$$= \frac{1}{P'Q'} \times \text{Difference of values of} \int (\mu + \delta\mu)\, ds$$

taken along the paths ST and SP', since along $P'T$ we have $\mu + \delta\mu = 1$.

Thus $$P'Q' . \delta\theta = \int_{RQ'} (\mu + \delta\mu)\, ds - \int_{SP'} (\mu + \delta\mu)\, ds,$$

the integral having the same values along the paths ST and RQ', since it is proportional to the time of propagation of the $(\mu + \delta\mu)$ wave.

Now $$\int_{RQ'} \mu ds = \int_{QQ'} \mu ds,$$

by the stationary property of $\int \mu ds$ (§ 3)

$$= \int_{PP'} \mu ds,$$

since the time of propagation of the μ wave is the same from any point on PQ to the corresponding point on $P'Q'$,

$$= \int_{SP'} \mu ds,$$

by the stationary property of $\int \mu ds$.

Thus we have

$$P'Q' . \delta\theta = \int_{RQ'} \delta\mu . ds - \int_{PP'} \delta\mu . ds$$

$$= \int_{QQ'} \delta\mu . ds - \int_{PP'} \delta\mu . ds,$$

to our degree of approximation.

If the prisms are all formed of the same variety of glass, this becomes

$$P'Q' . \delta\theta = \delta\mu \int_{QQ'} ds - \delta\mu \int_{PP'} ds,$$

where the integration is now to be taken only over those portions of the path which are inside the prisms, omitting the parts which are in air. Thus if t denote the difference of the lengths of path travelled in glass by the two sides of the beam, and if a denotes the breadth of the emergent beam, the last equation can be written

$$a\,\delta\theta = t\,\delta\mu.$$

Now if λ and $\lambda + \delta\mu$ denote the wave-lengths of the two radiations μ and $\mu + \delta\mu$, the resolving power of the spectroscope is (§ 35) $a\,\delta\theta/\delta\lambda$. Thus we have the result that *the resolving power of a prism spectroscope is*

$$t\,\frac{d\mu}{d\lambda},$$

where t denotes the difference of the lengths of path travelled in the glass of the prisms by the two sides of the beam, and $d\mu/d\lambda$ is the rate of change of refractive index with wave-length.

In the most usual case, one side of the beam passes through the refracting edges of the prisms, *i.e.* it does not travel any distance at all in the glass: and t then denotes practically the total length of those sides of the prisms which are opposite the refracting edges. Roughly speaking, one cm. of glass is required in order to resolve the yellow light of sodium into its two component radiations. It must, however, be remembered that the formula has been derived on the assumption that the slit is infinitely narrow: the small though measurable breadth of the slit diminishes the power of resolution.

In the early prism spectroscopes it was customary to use a large number of small prisms—often 12 or more—in order to obtain a high resolving power. The same end is now better attained by using a smaller number of prisms—generally not more than four—of much larger size. In the older arrangement the large dispersion caused a great separation of the different coloured beams even before their passage through the last prisms of the train, and consequently made it impossible for them to pass all together through the last prism: the full resolving power of the instrument was therefore only displayed over a very narrow range of the spectrum at once. This, though not a matter of much consequence in visual spectroscopes, where different parts of the spectrum can readily be brought to the centre of the field in turn, would be a serious defect if it were desired to photograph the spectrum. The loss of light by reflexion at the faces of the prisms was also much greater in the old than in the new type of spectroscope.

Printed in the United States
By Bookmasters